BIOTECHNOLOGY-4

INCLUDING

Recombinant DNA Technology
Environmental Biotechnology
Animal Cell Culture

AF651096

BIOTECHNOLOGY-4

INCLUDING

Recombinant DNA Technology
Environmental Biotechnology
Animal Cell Culture

S Mahesh

Former Professor
Department of Biotechnology
R V College of Engineering
Bangalore, Karnataka

A B Vedamurthy

Ex-Head, Department of Biotechnology
The Oxford College of Science
Bangalore, Karnataka

Publishing Globally

NEW AGE INTERNATIONAL (P) LIMITED, PUBLISHERS
LONDON • NEW DELHI • NAIROBI
Bangalore • Chennai • Cochin • Guwahati • Hyderabad • Kolkata • Lucknow • Mumbai
Visit us at www.newagepublishers.com

Copyright © 2003, New Age International (P) Ltd., Publishers
Published by New Age International (P) Ltd., Publishers
First Edition: 2003
Reprint: 2018

All rights reserved.
No part of this book may be reproduced in any form, by photostat, microfilm, xerography, or any other means, or incorporated into any information retrieval system, electronic or mechanical, without the written permission of the publisher.

GLOBAL OFFICES

- **New Delhi** **NEW AGE INTERNATIONAL (P) LIMITED, PUBLISHERS**
7/30 A, Daryaganj, New Delhi-110002, (INDIA)
Tel.: (011) 23253771, 23253472, **Telefax:** 23267437, 43551305
E-mail: contactus@newagepublishers.com • Visit us at www.newagepublishers.com
- **London** **NEW AGE INTERNATIONAL (UK) LTD.**
27 Old Gloucester Street, London, WC1N 3AX, UK
E-mail: info@newacademicscience.co.uk • Visit us at www.newacademicscience.co.uk
- **Nairobi** **NEW AGE GOLDEN (EAST AFRICA) LTD.**
Ground Floor, Westlands Arcade, Chiromo Road (Next to Naivas Supermarket)
Westlands, Nairobi, KENYA, **Tel.:** 00-254-713848772, 00-254-725700286
E-mail: kenya@newagepublishers.com

BRANCHES

- **Bangalore** 37/10, 8th Cross (Near Hanuman Temple), Azad Nagar, Chamarajpet, Bangalore- 560 018
Tel.: (080) 26756823, **Telefax:** 26756820, **E-mail: bangalore@newagepublishers.com**
- **Chennai** 26, Damodaran Street, T. Nagar, Chennai-600 017, **Tel.:** (044) 24353401
Telefax: 24351463, **E-mail: chennai@newagepublishers.com**
- **Cochin** CC-39/1016, Carrier Station Road, Ernakulam South, Cochin-682 016
Tel.: (0484) 2377303, **Telefax:** 4051304, **E-mail: cochin@newagepublishers.com**
- **Guwahati** Hemsen Complex, Mohd. Shah Road, Paltan Bazar, Near Starline Hotel
Guwahati-781 008, **Tel.:** (0361) 2513881, **Telefax:** 2543669
E-mail: guwahati@newagepublishers.com
- **Hyderabad** 105, 1st Floor, Madhiray Kaveri Tower, 3-2-19, Azam Jahi Road, Near Kumar Theater
Nimboliadda Kachiguda, Hyderabad-500 027, **Tel.:** (040) 24652456, **Telefax:** 24652457
E-mail: hyderabad@newagepublishers.com
- **Kolkata** RDB Chambers (Formerly Lotus Cinema) 106A, 1st Floor, S N Banerjee Road
Kolkata-700 014, **Tel.:** (033) 22273773, **Telefax:** 22275247
E-mail: kolkata@newagepublishers.com
- **Lucknow** 16-A, Jopling Road, Lucknow-226 001, **Tel.:** (0522) 2209578, 4045297, **Telefax:** 2204098
E-mail: lucknow@newagepublishers.com
- **Mumbai** 142C, Victor House, Ground Floor, N.M. Joshi Marg, Lower Parel, Mumbai-400 013
Tel.: (022) 24927869, **Telefax:** 24915415, **E-mail: mumbai@newagepublishers.com**
- **New Delhi** 22, Golden House, Daryaganj, New Delhi-110 002, **Tel.:** (011) 23262368, 23262370
Telefax: 43551305, **E-mail: sales@newagepublishers.com**

ISBN: 978-81-224-1442-4

C-17-07-10639

Printed in India at Star Print-O-Bind, Delhi.
Typeset at Pagitek, Delhi.

NEW AGE INTERNATIONAL (P) LIMITED, PUBLISHERS
7/30 A, Daryaganj, New Delhi-110002
Visit us at **www.newagepublishers.com**
(CIN: U74899DL1966PTC004618)

PREFACE

Biotechnology is one of the fastest emerging fields because of its importance in areas like agriculture, pharmaceutical industry, biochemical and biomedical engineering. The progress made in this field has a direct impact on mankind and his welfare. Biotechnology is not just any technique but also touches on the foundation of life.

There are innumerable contributions of recombinant of DNA technology that have been recognised in industry, agriculture and medicine. A substantial growth of biotechnology industry mainly stems from the release of recombinant DNA products in the market. Energy for cleaner environment and energy security is the slogan of the day. Environmental biotechnology helps to maintain cleaner environment and its significance on human health is exemplary. The overriding principle must be that the use of environmental biotechnology should be safe for both people and for the environment. Hence this field can make a major contribution in treating industrial, municipal and contaminated waste with heavy metal and oil spill. Animal cell culture technique can encompass to cell cloning, stem cell culture, and in the production of useful recombinant proteins. Production of cloned protein in animal cell culture is being considered as one of the most significant contributions in the field of biotechnology.

This book covers three important branches of biotechnology such as Recombinant DNA Technology, Environmental Biotechnology and Animal Cell Culture. These are highly interrelated fields in emerging areas of biotechnology. This book is simple, understandable for readers, particularly for students. This book is also useful for the students studying UGC vocational studies for biotechnology.

We express our gratitude to Mr. Saumya Gupta, M.D., New Age International Publishers, New Delhi for his acceptance in publishing this book.

S. MAHESH
A.B. VEDAMURTHY

Contents

Part A
Recombinant DNA Technology

1

What is Recombinant DNA Technology

Recombinant DNA technology is one of the most fascinating field in biotechnology. It stems from an understanding of functions of DNA and discoveries in molecular biology. The farmost technique in recombinant DNA technology is the transfer of a sequence of DNA (gene) from one organism to other organism. Hence, it is also called as "molecular cloning" or "gene cloning" or broadly "genetic engineering".

In microorganisms, persistent endeavour has been made to increase desirable gene pool by creation of mutants using chemical mutagens or high energy radiation. However, since there is no clear view on, which gene is mutated, the vast number of mutants are lethal in conventional breeding. Conjugation or sexual method of gene transfer is perhaps a relatively random process. Hence, one cannot totally rely on these traditional methods to increase novel set of gene pool.

By contrast, genetic engineering is especially useful when, one wishes to introduce a gene from higher organisms in bacteria or yeast. Since manipulation may result in large scale production of several pharmaceutically important compounds, the exploitation of poor microorganisms for the benefit of mankind have become the order of the day. We can tame and order microorganisms to do the job by recombinant DNA strategy.

1.1 WHY GENE CLONING?

There are several possible reasons for wishing to clone a particular piece of DNA. One might wish to introduce a gene of interest in bacterial cells that codes for valuable proteins in such a way that, the cells would produce large amount of proteins of interest. A tiny bacterium acts as a microbiological factory which helps us in producing large amount of insulin, growth hormones, etc. Before that it is quite but natural to introduce, "what is gene cloning"? In the simplest term "gene cloning" is essentially an insertion of a specific piece of foreign DNA into a cell, in such a way that the inserted DNA is able to replicate and pass onto daughter cells during cell division. The inserted DNA is known as "transgene" and any organism containing the transgene is called "Transgenic" or "Genetically modified organism".

The essential steps of gene cloning are as follows:

(a) Fish out a gene of interest (target gene) from the DNA of an organism.

(b) Insertion of the target gene into another piece of DNA called a "vector". The vector may be usually a bacterial plasmids or viral DNA or cosmids.

(c) The inserted DNA in vector in transferred back into bacterial cells.

(d) The presence of recombinant molecule in bacteria is screened and only transformed cells are selected.

(e) Expression of a cloned gene through controlled elements.

The successful story of transformation stems from the discovery of several key tools such as DNA cutting and DNA rejoining enzymes. The vehicles for efficient transport of our gene of interest to the target cells are called vectors. These are the perfect mediators to transfer gene of interest from one organism to another.

1.2 WHY DO WE HAVE TO CLONE A GENE!

The amazing advancement made in gene cloning has revolutionised the generation of new genetically modified organisms, including plants and animals.

Although conventional breeding has helped the scientist to generate novel characters in higher organisms such as, plants and animals, the perfect differential crossing combines the good features of the desired gene, unfortunately, the chances of the occurrence of such a perfect combination of genes tends to vanish in due course of time or meagre chances of combination of genes and untimely expression, were found to be major lacunae. In order to overcome several breeding barriers in microorganisms and plants and animals, a new line of invention has been developed in the field of modern biology. The clear understanding of genetic material of bacteria and viruses laid a foundation for recombinant DNA technology. The following information would undoubtedly justify the term "why do we have to clone a gene!".

1.2.1 Microorganisms

The exploitation of microorganisms since many centuries for fermented food products and beverages are well known. These are extensively used in industries for the production of useful organic compounds. The outcome of gene cloning technique has further advanced in exploiting microorganisms for the production of many recombinant products. The major use of recombinant DNA technology using microorganisms today is, the production of proteins. At present, the genes for over 200 different proteins that are potential human therapeutic agents have been cloned and are currently being tested for the treatment of various human diseases. Insulin was the first recombinant drug produced commercially in the bacterium *E. coli*. In addition, several vaccines, interferons and human growth hormones have also been produced through recombinant DNA strategy.

1.2.2 Plants

Plant genetic engineers transfer genes from one plant to another. However, these engineering fantasies targetted microbial genes and transferred them to plant system to obtain disease resistant and insect resistant plants. By using genetic engineering strategies, these enthusiastic engineers increased the shelf life of fruits and cut flowers. The production of edible vaccines, monoclonal antibodies, flavouring agents are some of the additional list of achievements in genetic engineering of plants.

1.2.3 Animals

Genetic engineering of animals is restricted to the extent of subdividing cluster of cells at the early stages of embryo development. The lack of totipotency is one of the main obstacle in animal system. However, successful cloning of animals has increased the enthusiasm among scientists to clone human being. This may overcome several lacunae in animal system.

Some successful attempts have already been made to obtain expression of foreign genes in whole animals. Transgenic animals can be a model system for studying gene therapy which is aimed at several unsolved human genetic diseases. The livestock improvement by increasing the growth rate, milk yield, are some of the key areas in genetic engineering. Milk producing mammary gland in animals especially cow and buffaloes can be an ideal choice for the production of recombinant product.

QUESTIONS

1. What is gene cloning? Why do we have to clone a gene?
2. Explain the steps involved in gene cloning.
3. Describe the process of creating recombinant DNA molecule.

2

Tools in Recombinant DNA Technology

The engineering of genetic material was one of the prime goal for modern biotechnologists in the early 1970's. There was no method available for cutting and pasting of DNA. With the advancement in the new inventions several scientific tools became available for genetic manipulation of DNA. One of the most remarkable achievements in the emergence of genetic engineering is the discovery of restriction endonucleases and their specific functions in bacteria. The reality of restriction enzyme as a molecular scissor contributed significantly towards gene splicing and analysis of cloned genes.

The inventions of different types of vectors and their role in cloning of specific gene of interest have been found to be exemplary. The utility of bacterial plasmids and viruses in accommodating foreign gene of interest and their efficient expression gained significant momentum in Recombinant DNA techniques.

2.1 RESTRICTION ENDONUCLEASES

Restriction enzymes or Restriction endonucleases belong to a class of bacterial enzymes that cut DNA molecules internally at specific recognition site. The discovery of restriction enzymes was made by Hamilton Smith in 1970, from John Hopkins University, in the bacterium *Haemophilus influenzae*. Since then several hundreds of restriction enzymes have been isolated from several bacterial strains and today over 150 different DNA cleavage sites have been established.

Restriction enzymes are popularly known as "Molecular Scissors" and mainly classified as "restriction endonucleases and exonucleases". Restriction endonucleases cut the DNA backbone at an internal site on the molecule. All restriction endonucleases have specific recognition sequence within the

DNA molecule. These sequences generally fall between four to six base pairs in length. The restriction recognition sequence are often known as "palindromic sequences" i.e., the two strands are identical when read both in forward and backward direction, in this region. Each individual enzyme recognises only one specific sequence. However, the same specific sequence is recognised by two or more different enzymes. Such types of enzymes are called "isochizomers".

2.1.1 Types of Endonucleases

Restriction endonuclease can be divided into type I, II and III (see Table 1). The type I enzyme recognises specific sequence but cut the DNA molecule far away from the recognition sequence. The cleavage takes place about 1000 base pairs away from recognition site. Hence, the use of class I enzyme in genetic engineering is restricted. Class I requires Mg^+ and ATP as cofactors. On the other hand, the type II restriction enzymes are able to recognise specific sequence and cut the DNA molecule at the same specific site or within the specific sequence. These class of enzymes are the major choice in genetic engineering work. Majority of type II restriction enzymes recognise tetra, penta or hexa and sometimes even hepta nucleotide sequence. They require Mg^{2+} for restriction digestion. The classical example for class II type restriction enzyme is EcoR I.

2.1.2 Nomenclature:

In addition to EcoR I, hundreds of other type II restriction enzymes have been isolated from bacteria. The following points are essentially followed in the naming protocol for restriction endonucleases.

(1) The three letter code is given for each restriction endonucleases.
(2) The first letter code is always a genus name and written in capital.
(3) The subsequent letters are represented by species name.
(4) The next letter designates the strain, and roman numbers indicate different endonucleases from the same organism. Example EcoR I, EcoR II.

Table 2.1. Types of restriction endonucleases and their target sequences

	Name of the enzyme	Target sequence	Types of DNA end
1.	Bam HI *(Bacillus amyloliquefaciens)*	G ↓ GATCC	5' phosphate extn.
2.	EcoR I *(Escherichia coli)*	G ↓ AATTC	5' phosphate extn.
3.	Hae III *(Haemophilus aegyptius)*	GG ↓ CC	Blunt end
4.	Sma I *(Serratia marcescence)*	CCC ↓ GGG	Blunt end
5.	Pst I *(Providentia stuartii)*	CTGCA ↓ G	3' hydroxyl extn.
6.	Hind III *(Haemophilus influence)*	A ↓ AGCTT	5' phosphate extn.
7.	Alu I *(Arthrobacter luteus)*	A ↓ GCT	5' phosphate extn.

2.1.3 Chemistry of Restriction Enzymes:

Structurally most of the restriction enzymes are made up of two equal subunits, with molecular weight of 20,000-25,000 daltons or single polypeptide with molecular weight of 30,000-35,000 daltons.

2.1.4 Restriction Enzymes Generates Staggered and Blunt Ends:

Restriction enzymes after digestion with DNA create staggered end or sometimes blunt end depending on the type. The restriction enzyme Hae III cut the backbone of both the strands of DNA within recognition site, as a result two new fragments have square or blunt ends.

Example:

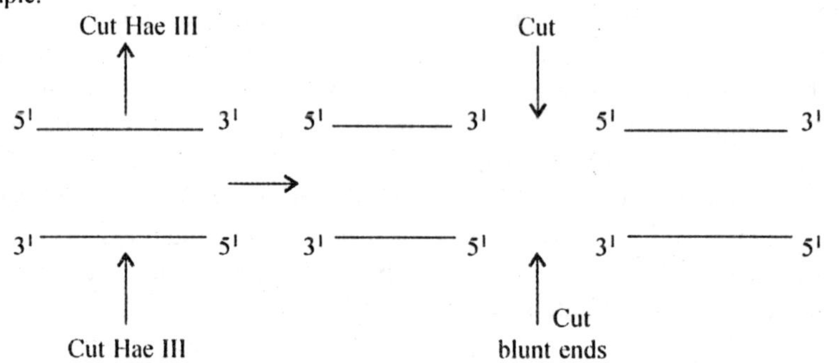

Some restriction enzymes such as EcoR I does not cut the DNA in a blunt-ended fashion. Instead, it generates sticky end or protruded end (5^1 or 3^1 phosphate extension). The *Sticky end* is also known as *Cohesive end.*

Example:

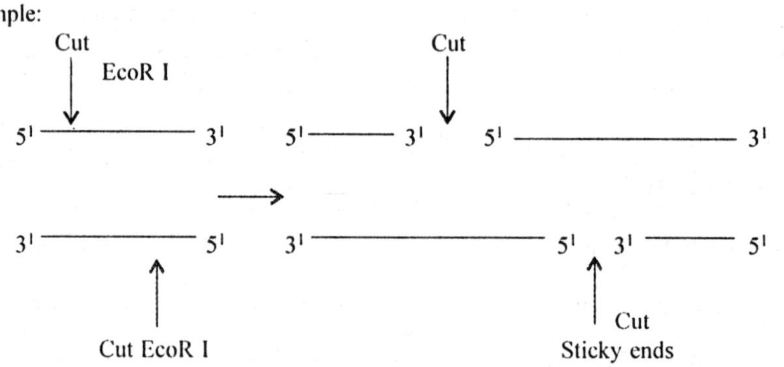

Restriction enzymes which produces cohesive end or sticky ends are useful in genetic engineering work.

2.1.5 Purification:

The viscosity change, loss of biological activity are some of the parameters followed in the detection of enzymes. The phospho cellulose, sepharose and cibacrone are widely used in chromatographic technique to separate restriction enzymes.

2.2 DNA LIGASE:

It acts as a key player in recombinant DNA technology as a "Molecular Suture". The main source of this enzyme is the T_4 phage virus. DNA Ligase can stitch (join) two different DNA pieces. The joining process requires ATP as it derives energy to construct the bond. Normally the process is carried out at 4°C in order to reduce the kinetic energy of the molecule. This enzyme, when joining two DNA pieces, forms a covalent bond between the 5^1 phosphoryl of one strand and 3^1 hydroxyl of the adjacent strand. Hence, it catalyses the end to end joining of DNA duplex at the base paired end. During sealing process, two phospho diester bonds are formed by T_4 DNA ligase. The association of fragments is stabilised by the formation of 3^1 to 5^1 phospho diester linkage between cohesive ends. The blunt ends produced by certain restriction enzymes may hinder the sealing process. Hence, it is estimated to use ligase at a very high concentration to complete sealing process (Fig. 2.1).

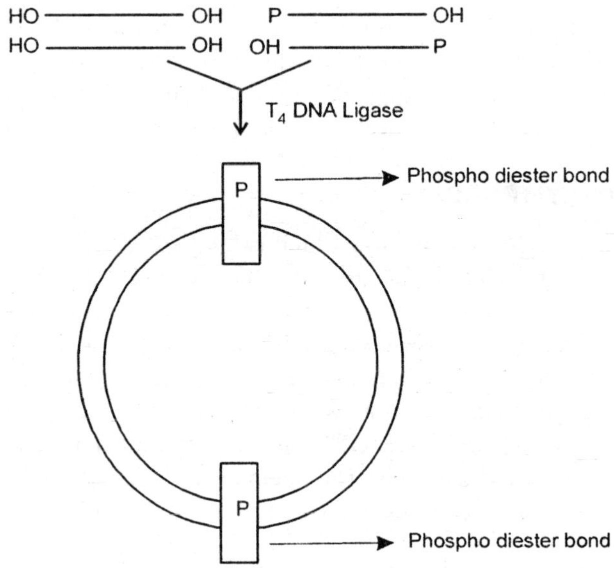

Fig. 2.1. Joining of two DNA fragments by DNA ligase.

2.3 ALKALINE PHOSPHATASE:

Alkaline phosphatase is a glycoprotein with two identical sub units. It catalyses the removal of 5^1 terminal phosphate residue from nucleic acids. Alkaline phosphatases are used to reduce unwanted ligation of cleaved plasmids during the preparation of recombinant DNA molecule i.e. dephosphorylation of vector DNA will prevent recircularisation of vector in cloning procedure. This enzyme is also used for mapping and finger printing studies.

2.4 DNA POLYMERASE I:

DNA polymerase is a holoenzyme. The larger fragment of this enzyme is known as Klenow fragment. In genetic engineering work, DNA polymerase is widely employed in the synthesis of double strand of cDNA and also in dideoxy method of DNA sequencing.

2.5 RNases:

RNases cleaves phosphodiester bond between the two adjacent nucleotides. RNase cleaves the bond next to uracil and guanine. The main source of this enzyme is *Aspergillus* sp. and bovine pancreas. The RNase H is widely used in cDNA preparation. It removes mRNA from RNA-DNA hybrid. RNase is invariably used in DNA purification where it will eliminate contaminated RNA and set free DNA. This enzyme is also used to remove poly A tail.

2.6 DNases:

DNases can digest both the strands of DNA. It can hydrolyse each DNA strand independently in presence of Mg^{+} ion. DNases are used in RNA purification where it can eliminate the contaminated DNA and set free RNA, besides its wide utility in foot printing and nick translations.

2.7 Taq POLYMERASE:

This enzyme is DNA polymerase which exhibits its activity at temperature optimum of 72°C. This highly thermostable enzyme is extracted from the bacterium *Thermus aquaticus* living in hot springs at a temperature of more than 75°C. Taq polymerase consists of single polypeptide chain with a molecular weight of 90,000 daltons. The main application of Taq polymerase is in polymerase chain reaction, where it is used to extend primers and completes the synthesis of new DNA strand. The specificity and sensitivity of the PCR is also improvised by Taq polymerase. Fidelity of DNA synthesis by Taq polymerase determines the accuracy of PCR amplification.

2.8 REVERSE TRANSCRIPTASE:

It is an RNA dependent DNA polymerase obtained from avian myeloblast virus. This enzyme requires DNA primers complementary to the RNA template. It consists of two important sub units. It exhibits a variety of activities such as RNA directed DNA polymerase, DNA-dependent DNA polymerase etc. In genetic engineering, reverse transcriptase is well known in cDNA preparation, where it converts mRNA to single strand DNA.

2.9 MOLECULAR CLONING VECTORS:

As a molecular tool various cloning vectors are currently being used in genetic engineering. These are plasmid cloning vectors, phagemids and cosmids.

2.9.1 Plasmid Cloning Vector (*For E. coli*):

Plasmids are self replicating, double stranded circular DNA molecules that are present in bacteria. The plasmids perform several functions including resistance to antibiotics. The size of plasmids ranges from 1 to more than 500 kilo base pairs. The number of plasmids ranges from 10-100 per cell and they are known as "high-copy-number plasmids or relaxed plasmids". However, some host cell contains 1 to 4 copies and are called "low-copy-number plasmids or stringent plasmids". Plasmids itself carries number of genes for conjugation, selectable markers and specific antibiotic resistance. Following are some of the plasmid based cloning vectors.

2.9.1.1 pBR 322: It is undoubtedly the ideal cloning vector, commonly employed in genetic engineering. It is derived from *E. coli* plasmid and engineered into a cloning vector. pBR 322 plasmid consists of 4,363 base pairs and carries genes which confirms resistance for two antibiotics, namely, tetracycline (Tet^r) and ampicillin (Amp^r). The gene for tetracycline contains unique recognition sites for Hind III, Bam HI and Sal I. Similarly ampicillin resistance gene contains pst I site (Fig. 2.2).

2.9.1.2 Nomenclature: In pBR 322 the first letter p denotes plasmid and BR indicates the names of two scientists, Bolivar and Rodriguez, who constructed this plasmid. 322 is a number given to differentiate other plasmids.

2.9.1.3 Screening of pBR322: The presence of two antibiotic resistance genes are beneficial during selection and screening process. The screening is actually done in two steps. First the transformed cells are grown in ampicillin containing medium. Only transformed cells containing ampicillin resistance gene are able to grow. However, non transformed cells are sensitive to ampicillin and do not grow. In the subsequent step the cells that are grown on ampicillin media are transferred to a tetracyclin containing media. In transformed cells the tetracyclin gene is disrupted or damaged during insertion of target DNA as it lies within this gene. The cells that do not grow on tetracyclin medium are sensitive to tetracyclin and actually carry a cloned DNA.

2.9.1.4 Significance of pBR 322: The laboratory designed pBR 322 is smaller than a natural plasmid. Generally, reduction in the size of the plasmid can minimise the damage and increase the efficiency of transformation. The presence of antibiotic resistance genes is an added advantage during screening procedure. The instability of this plasmid in continuous and batch culture, presence of low copy number per cell are some of the disadvantages of this designed plasmid.

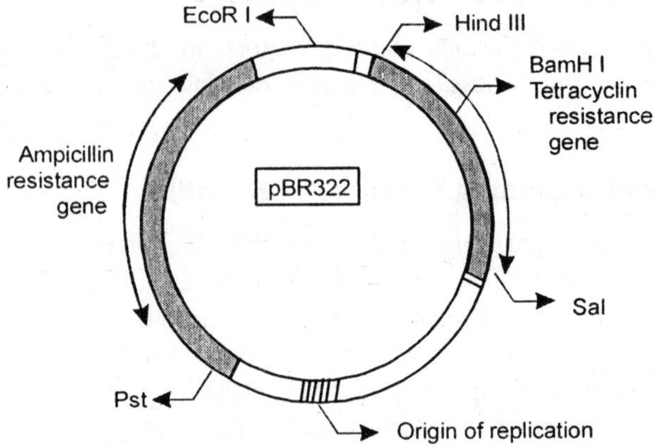

Fig. 2.2. Diagram showing the plasmid pBR 322.

2.9.2 pUC 19

The plasmid pUC has 2,680 base pairs. It carries gene for ampicillin-resistance, β-galactosidase (lac Z), lac I. The lac Z gene contains short sequence for multiple unique cloning sites i.e., EcoR I, Sac I, Xma I, Sma I, BamH I, Pst I, Hind III etc. and the origin of replication. The main advantages of pUC in genetic engineering is easy selection in screening process. It has definite advantage over time consuming selection procedures followed in pBR 322.

2.9.2.1 Nomenclature: In pUC 19, the first letter 'p' denotes plasmid. The letters U and C are referred to the University of California, where it was constructed by J. Messing and J. Vieira. The number 19 distinguishes from other plasmids (Fig. 2.3).

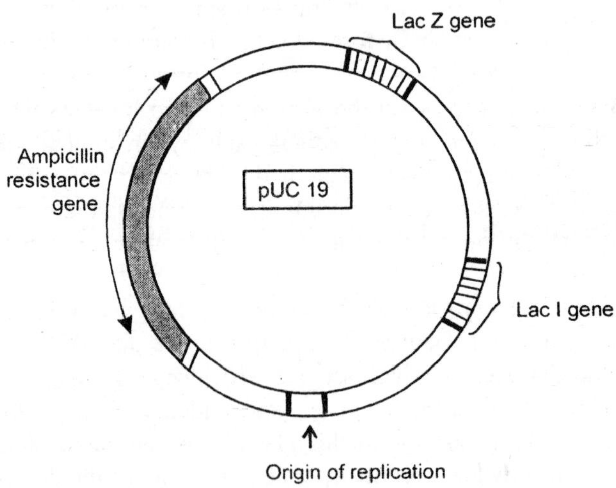

Fig. 2.3. Diagram showing plasmid pUC 19.

2.9.2.2 Significance of pUC 19: The significant feature of pUC 19 is that it can exist as high copy number i.e., upto 500 per bacterial cell. It is easy for isolation and contains a variety of restriction enzyme recognition sites. The lac Z gene is under the control of lac I promoter. This mechanism is used in the production of certain foreign fusion proteins.

Table 2.2. Comparative difference between pBR 322 and pUC 19

	pBR 322	pUC 19
Size	4,363 bp	2,680 bp
Unique cloning sites	Hind III, Bam HI, Sal I, Pst I	EcoR I, Sac I, Xma I, Sma I, Bam HI, Pst I, Hind III
Type of gene	Ampicillin, Tetracycline	Ampicillin resistance gene, lac Z and lac I gene

2.10 VIRUS BASED VECTORS

2.10.1 Bacteriophage λ Vectors:

Although plasmid based vectors are efficient in transformation, they can carry only up to 10 kb of inserted foreign DNA. However, in genetic engineering it is essential to create a vector capable of carrying large pieces of DNA. To meet this demand, the *E. coli* virus – Bacteriophage Lambda (λ) has been the foremost choice. Lambda virus can accommodate large pieces of inserted DNA.

The bacteriophage λ approaches two types of replication inside the host cell. During lytic cycle it can attach and enter the bacterial cell followed by replication. After a short period of time, newly formed phages are released after the lysis of bacteria. In lysogenic cycle λ DNA enters bacterial DNA and become silent for a period of time. During favourable conditions λ DNA can be excised and enter a lytic cycle.

The bacteriophage λ DNA consists of 50 kb, of which 20 kb, is responsible for integration and excision during lysogenic cycle. This middle region of λ DNA of about 20 kb can be removed and replaced with 20 kb of foreign DNA. During Vector construction the source DNA is cut with restriction enzymes and DNA sequence of 20 kb is excised. Meanwhile middle portion of λ DNA of 20 kb is cut and removed. Later λ DNA and source DNA are mixed and incubated in presence of T_4 DNA ligase. These molecules are packaged into bacteriophage λ heads as infective particles are formed after the addition of tail assemblies. The recombinant bacteriophage λ are maintained by continual growth on fresh *E. coli* (Fig. 2.4).

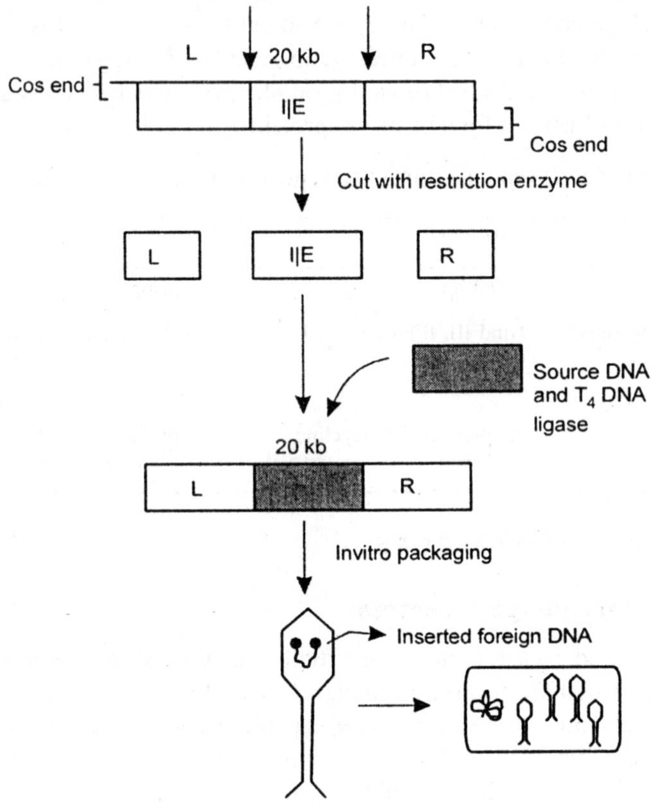

Fig. 2.4. Construction of λ phage vector.

2.10.2 M13 Bacteriophage Vector:

M13 is a single-stranded DNA containing phage virus. It is filamentous in nature. The protein coat contains three kinds of capsomers. These phages

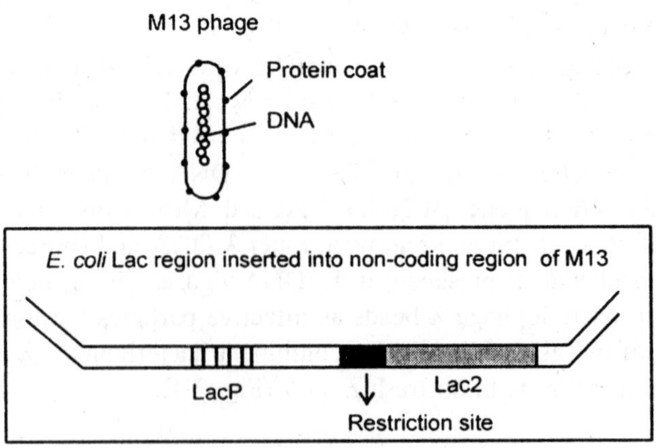

Fig. 2.5. Modification of M13 vector.

infect cells by adsorbing with the aid of F. pili. These infect only F^+ cells not F^- cells. The DNA of wild type M13 comprises approximately 6,402 bases. The presence of 10 genes are essential for replication of the phages (Fig. 2.5).

In M13 cloning vector the 507 nucleotide intergenic sequence (IS) is modified to introduce additional restriction site. Intergenic sequence of wild type is the only region available for modification. The gene lac Z is introduced into the intergenic sequence (IS) to get M13 vector. In the screening procedure the introduced lac Z gene produces blue plaques on X gal agar plates.

2.11 COSMIDS:

Cosmids are engineered cloning vectors and can carry about 40 kb of foreign DNA. Cosmids combine unique properties of plasmid and bacteriophage Lambda DNA. It can be easily maintained in *E. coli* cells. Cosmids has two cos sites derived from λ phage and tetracycline resistance gene from plasmid. It has a multiple cloning sequence with six restriction sites. The source DNA is cut with restriction endonucleases and DNA size of about 40 kb is isolated. The plasmid DNA also cuts with enzyme. Subsequently source DNA and plasmid DNA are mixed and incubated in presence of T_4 DNA ligase. After ligation, 40 kb of foreign DNA is inserted into the cosmid. These molecules are packaged into bacteriophage λ heads invivo and infection particle containing foreign DNA can be obtained after addition of tail assemblies (Figs. 2.6 and 2.7).

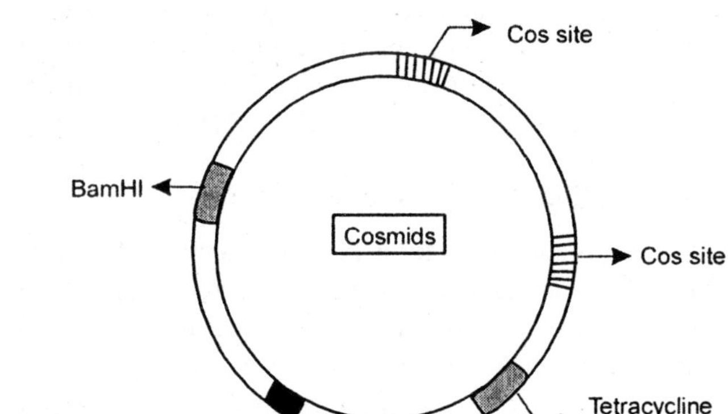

Fig. 2.6. Diagram of cosmid.

2.12 SHUTTLE VECTORS

These are the type of vectors which are capable of replicating in different host systems. As shuttle vectors carry different origin of replication, they can be easily maintained in different host systems. Shuttle vectors can be maintained both in *E. coli* and yeast cells. One of the reason for selecting eukaryotic yeast cell is that the lacunae of post translational modification. Yeast cells of eukaryotes are better equipped for post translational modification. This is much needed for the production of eukaryotic proteins. Hence, most eukaryotic vectors are designed to be shuttle vectors.

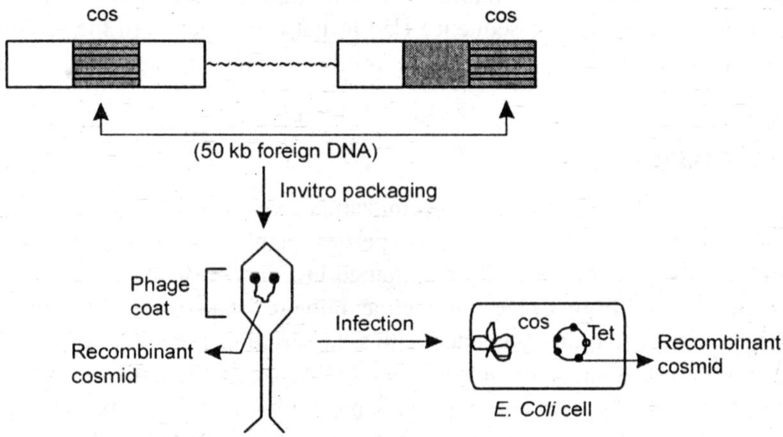

Fig. 2.7. A cosmid cloning system.

2.12.1 Yeast Episomal Plasmids

Yeast episomal plasmids (YEP) are shuttle vectors. They can replicate in dual host cells, i.e., *E. coli* and yeast. In other words, they carry origin of replication for *E. coli* and yeast cells as well as marker gene. One of the example of yeast episomal plasmids is PJDB 219. It is an artificial plasmid found in strains of *Saccharomyces cerevisiae* with copy number upto 250.

2.12.2 Yeast Artificial Chromosome (YAC)

In the laboratory, this shuttle vector is designed to clone more than 100 kb of DNA, which is then maintained as a separate chromosome in the host yeast cell. The YAC mimics a chromosome because it contains a unique sequence that acts as an origin of DNA replication.

The YAC plasmid has an *E. coli* selectable marker gene, an origin of replication and yeast DNA sequences such as URA, CEN, TRP and ARS. URA gene for uracil biosynthesis. TRP for tryptophan, CEN provides centromere function and ARS for yeast autonomous replicating sequence.

2.13 VIRAL VECTORS FOR ANIMALS

2.13.1 SV_{40} or Simian 40, virus

It contains single double stranded circular chromosomes. It normally infects monkey cells and induce tumour. The genes of SV_{40} chromosomes can be divided into two sets. One is known as "early genes", which encodes for tumour antigen and another is "late genes", which encodes for viral coat protein.

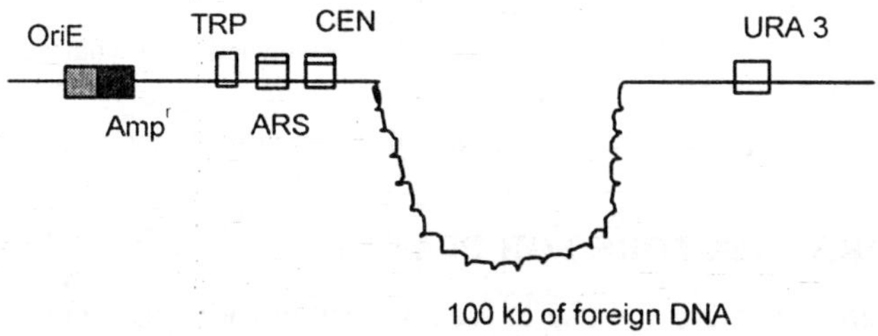

Fig. 2.8. Yeast artificial cloning system.

Simian virus contains small genomes. It cannot accommodate large pieces of foreign DNA. Hence, an essential portion of viral genome is removed. The defective SV_{40} can no longer cause infection. Therefore with the help of helper virus, defective virus containing foreign gene can perform infection and participate in gene transfer for mammalian cells.

2.13.2 Retrovirus

Some of the retroviruses have been modified to act as viral vector to introduce foreign DNA into mammalian cells. When a retrovirus enters the host cell, its RNA is converted into double-stranded DNA by reverse transcriptase and is then integrated into host genome. The LTR (Long terminal Repeats) sequences are required for integration of retroviral DNA into the host chromosome. The psi (ϕ) sequence is required to package viral RNA in viral particles. The gag, pol genes of the retroviral genome can be replaced with foreign DNA. This results in the formation of recombinant defective retroviral DNA. This recombinant DNA lacks the genes required for retroviral replication and assembly of viral particles. To assemble viruses with recombinant, it must be introduced into cells in tissue culture that are infected with helper viruses. Within the cells, recombinant DNA is transcribed and RNA is packaged. These viral particles contain only recombinant viral RNA and can act as vectors to introduce this RNA into target cells.

2.13.3 Baculovirus

Baculoviruses infect invertebrates including many insects. In baculovirus *Autographa californica,* a multiple nuclear polyhedrosis virus (AcMnPv), which infects insect cells, has been developed as a eukaryotic expression vector, with a promising result. This system places a cloned gene into a site in an AcMnPv virus that, in cultured insect cells, produces large quantities of recombinant protein. In this system there is exceptionally high gene expression, driven by AcMnPv promoter late in the infection cycle (Fig. 2.9).

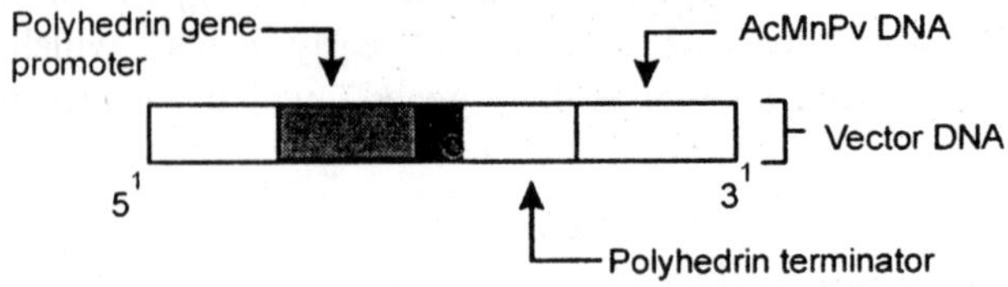

Fig. 2.9. Organisation of *Baculovirus* vector.

2.14 VIRAL VECTORS FOR PLANTS

Viruses are attractive for use as vectors for introducing genes into plants. The main advantage is that they can distribute their own genome throughout as infected plant. Experiments have been performed by using geminiviruses to assess their potential as vectors.

A recombinant version of DNA of tomato golden mosaic virus was made with an NPT II gene in place of the viral coat protein gene. This fragment is introduced into *Agrobacterium* system. When those bacteria were injected directly into the stems of transgenic tobacco, spreading of viral DNA was noticed as NPT II gene is expressed.

2.14.1 Cauliflower Mosaic virus (CaMv)

It is a double stranded DNA virus. This virus infects plants of the family Cruciferae. The naked CaMv DNA is capable of infecting host cells. Infection can be initiated by inoculating the host plant with DNA. The genes of CaMv are closely packaged and it also lacks essential sequences that could be replaced by foreign genes. A small foreign DNA, having bacterial dihydrofolate reductase gene has been successfully expressed in plants.

QUESTIONS

1. What are type II restriction endonucleases and why are they important for recombinant DNA technology?
2. What are Molecular Scissors? Explain their role in genetic engineering. (Nov. 1997 BU, April 2000)
3. Write an account of molecular vectors employed in genetic engineering. (BU April 1996, Nov. 1999)
4. Give a brief account of bacterial vectors used in genetic engineering. (BU Nov. 1997)
5. Give a brief account of Gene isolation and cloning. (BU May 1996)
6. Elucidate methods of creating Recombinant DNA molecule (BU April 2000)

7. Describe the use of pBR 322 as a cloning vector. Add a note on its special feature.
8. Why would you use a plasmid, bacteriophage λ or cosmid as a cloning vector?
9. How do you modify λ phage virus into a cloning vector.
10. What are shuttle vectors. Explain their role in genetic engineering.
11. Highlight the distinguishing features of plasmids and cosmids as cloning vectors.
12. Describe viral based cloning vectors for bacteria, animals and plants.

3

Purification of DNA from Living Cells

Purification of DNA from intact living cells lies in the heart of molecular biology. It involves complete elimination of carbohydrates, proteins, lipids, even RNA and set free only DNA. Several hurdles are associated with the extraction of DNA. Some of the main hurdles are contamination of DNases enzyme which can be overcome by using liquid nitrogen. The laboratory contamination of DNases can be minimised by sterilisation of all the glasswares. In order to extract DNA from a bacterial cell, it is necessary to treat the cells with an enzyme called "lysozyme"' in order to weaken the bacterial cell wall. The weakened bacterial cell wall is subsequently treated with an anionic detergent such as sodium dodecyl sulphate (SDS).

Eukaryotic cells such as plant cells possess a rigid cell wall. Hence, it is indispensable to treat cells with an enzyme cellulase or using abrasive to cell wall weakening chemicals are most beneficial. Since animal cells do not possess cell wall, but cell membrane, to isolate the DNA, the cells are treated with an anionic detergent, such as SDS (sodium dodecyl sulphate) to solubilize the cell membrane. The DNA in the lysate can be extracted by suitable extraction procedure.

3.1 EXTRACTION OF BACTERIAL DNA:

Step 1:

A known weight of packed bacterial cells is added to an extraction medium consisting of saline EDTA. The high salinity in the extraction medium (NaCl) helps to prevent strand separation by providing a counter ion shield on the double helix of DNA. The presence of EDTA in the medium ensures chelation of metal ions. These ions are necessary for the activity of the enzyme DNase and thereby inhibits the activity of the enzyme.

In order to weaken bacterial cell wall, small amount of lysozyme is supplemented and the suspension is kept at 37°C for 20 min in water bath. The weakened cell walls are further ruptured by the addition of 10% SDS. After initial incubation at 60°C for 10 min, it is then cooled on ice to bring to room temperature.

The protein present is denatured by the addition of sodium chlorate solution and the DNA is extracted by an organic solvent mixture containing chloroform and isoamyl alcohol. Chloroform is the most important solvent for extraction, as it can denature proteins. The presence of isoamyl alcohol acts as an anti-foaming agent. DNA that is present is then purified by precipitation with ethanol and later with isopropyl alcohol.

Step 2: First ethanol precipitation

The resulting emulsion obtained from step 1 protocol is centrifuged at 10,000 rpm. The upper aqueous layer is removed and kept in a beaker and the solution is overlayered with 80% ethanol which is added slowly along the sides of the beaker. The two layers are shaken with the help of a glass rod and the DNA fibres are wound out on the rod or more precisely spooling on the glass rod. It is very essential to remove the solvent by pressing the DNA mass against sides of the glass container. Finally fibrous DNA is blotted with filter paper.

Step 3: RNAase treatment and second ethanol precipitation:

The DNA is dissolved by saline sodium solution. The presence of sodium citrate in saline solution can also check the activity of DNases. Hence, incubation is required at 37°C for 30 min with occasional stirring. The incubation with the enzyme RNAse ensures the degradation of contaminated RNA and is kept at minimum. The RNase reaction is stopped by sodium chlorate solution followed by the addition of chloroform and isoamyl alcohol. The DNA is precipitated again after solvent removal by 80% ethanol and is spooled out by glass rod, purified and can be stored at 4°C.

Step 4: Isopropanol precipitation:

In order to precipitate residual DNA more amount of isopropyl alcohol is added. Precipitated DNA is removed, blotted with filter paper, redissolved in sodium saline citrate and stored at 4°C.

Step 5: Assay for nucleic acids

A suitable aliquot from the ethanol precipitate is assayed by spectral analysis. The UV absorption spectrum for DNA is obtained and the concentration of DNA is calculated by presuming that 1mg/ml of DNA gives an absorbance of 2.0 at 260 nm, in a 1 cm cuvette. The ratio of $E_{260} \times E_{280}$ should be between 1.8-1.9 for a pure DNA. A value less than 1.75 indicates protein contamination.

3.2 EXTRACTION OF PLASMID DNA:

Purification of plasmids from bacteria involves, culture of cells in liquid medium, harvested and a cell extract is prepared. The extract is deprotenised, the RNA is removed and the DNA is concentrated by ethanol precipitation. In a plasmid preparation, it is necessary to separate the plasmid DNA from the large amount of bacterial chromosomal DNA.

3.2.1 Separation on the Basis of Size:

The cells are lysed under very carefully controlled conditions, the DNA fragments are still very large, much larger than the plasmids and can be removed with cell debris by centrifugation. Cells are treated with EDTA and lysozyme. Lysis is induced by adding a non ionic detergent Triton X-100. This method causes little breakage in the bacterial DNA. After centrifugation, cleared lysate consists almost entirely of plasmid DNA.

3.2.2 Separation on the Basis of Conformation:

The conformational differences between plasmids and bacterial DNA can be used to separate the two types of DNA. Most plasmids exist in cell as supercoiled molecules. supercoiled molecules can be easily separated from non-supercoiled DNA. Two methods are used. In one method i.e., alkalinine denaturation, there is a narrow PH range at which non-super coiled DNA is denatured, whereas supercoiled DNA are not Sodium hydroxide is added to the cell extract. The PH is adjusted to 12 to 12.5, then the hydrogen bonding in supercoiled DNA is broken causing the double helix to unwind and two chains are separated. If acid is added now, the denatured bacterial DNA plasmids will reaggregate into mass. The insoluble network can be pelleted by centrifugation leaving plasmid DNA in the supernatant. RNA and protein can be removed by centrifugation steps.

3.3 ISOLATION OF DNA FROM PLANT TISSUE

Extraction of DNA from plant is more difficult than from bacterial system. Plant cell contains a variety of secondary metabolites such as phenols and flavanoids. These metabolites can interfere during extraction of DNA. Although several experimental protocols are available for plant DNA extraction, only general outlines of extraction is described as follows:

Step 1: Required amount of plant tissues are powdered quickly by liquid nitrogen. Alternately air dried tissues can be used. Sometimes oven dried or heat dried tissues at 40°C for 48 hr are beneficial and yield good amount of DNA.

The powdered sample is transferred to extraction buffer and incubated at 65°C for 1 hr. Generally extraction buffers contain chelating agents such as EDTA. The rigid plant cell wall is weakened by the presence of number of detergents such as SDS and CTAB (Cetyl Tri Amino Bromine). High ionic strength of buffer provided by NaCl facilitate dissociation of histone proteins. Most of the extraction buffer contains mercaptaethanol to cleave disulphide linkage in proteins.

Step 2: After incubation, solution is cooled to room temperature, chloroform is added and the suspension is subjected for centrifugation at 10,000 rpm for 10 min. Chloroform and concentrated phenols are used to denature proteins.

Step 3: The aqueous layer is removed after centrifugation and transferred into a clean beaker.

Step 4: Nearly 2/3 volume of isopropyl alcohol or ice cold ethanol is added slowly on the sides of glass walls to precipitate DNA. Glass rod is used to spool out DNA.

Step 5: The blotted DNA is dissolved in TE buffer and kept at 4°C till further use.

Step 6: The DNA dissolved in TE buffer is sedimented and subjected for RNase treatment to check RNA contamination.

Step 7: The quality of DNA is assayed by UV spectrophotometer at 260 nm.

3.4 ISOLATION OF DNA FROM ANIMAL TISSUE

Isolation of DNA from animal source is comparatively flexible as they do not possess the rigid cell wall and interfering secondary metabolites. Mostly animal DNA can be extracted by high salt or phenol extraction method. The chromatin proteins an dissociated by sodium dodecyl sulphate (SDS). Proteins are denatured by phenol-chloroform treatment. The general method of DNA isolation of animal tissue is as follows:

Step 1: Appropriate weight of animal tissue is homogenised in saline EDTA.

Step 2: Add SDS and phenol, shake well and centrifuge at least 10,000 rpm.

Step 3: To the aqueous layer chloroform and isoamyl alcohol is added.

Step 4: Add high conc of 80% phenol to denature proteins.

Step 5: After sedimentation of proteins, aqueous phase containing nucleic acid is taken and DNA is precipitated using ethanol and spooled out by glass rod.

QUESTIONS

1. Explain how DNA can be purified from bacteria.

2. Explain the methods of isolating DNA from living tissues.
3. Compare and contrast DNA isolation method between bacterial and plant system.
4. Explain the sequential steps of bacterial DNA isolation.

(BU April 2001)

4

Techniques in Recombinant DNA Technology

4.1 ELECTROPHORESIS

Electrophoresis is a separating technique where charged biomolecules such as proteins and nucleic acids, migrate in response to electric field.

Principles of electrophoresis

In electrophoresis the rate of migration in electric field depends on the field, net charge and size of the molecules. In addition, ionic strength, viscosity and temperature of the medium would also influence the rate of migration. Electrophoresis is a simple rapid and novel technique commonly employed in the field of biochemistry, molecular biology and physiology. Electrophoresis is classified into various types as follows.

4.1.1 Polyacrylamide Gel Electrophoresis (PAGE)

Polyacrylamide gel electrophoresis, commonly known as "PAGE" is widely used for the separation of proteins and low molecular weight DNA. In order to separate large number of samples under uniform electrophoretic condition vertical slab gel is extensively used.

4.1.2 Requirement

The essential requirements for electrophoresis are (a) Gel casting unit; (b) separation tank; (c) acrylamide gel; (d) buffer and (e) stains.

Polyacrylamide gel is prepared by mixing acrylamide and bis-acrylamide followed by addition of N, N, N′, N′-tetramethylethylenediamine (TEMED) (1%) and Ammonium per sulphate (10%). The bisacrylamide acts as a cross linking agent and polymerisation process is initiated by catalyst ammonium per sulphate and TEMED. The gel can be prepared between the two sandwiched glass plates. Initially separating gel is poured in between the glass plates, over which stacking gel is layered. In order to cast wells, comb is inserted before the gel gets polymerised.

The polymerised gel is removed and placed in the buffer filled separating tank. Different protein samples are mixed with tracking dye bromophenol blue and sucrose and are applied in the wells. The power pack with an output upto 500 V is switched on to supply direct current. Generally, 100-150 volts are given for protein separation (Fig. 4.1).

4.1.2.1 Staining

In order to visualise proteins, the gel has to be treated with suitable dye which produces stable coloured compound. Proteins separated on the gel are frequently stained with cooamassie brilliant blue which imparts bluish colour bands.

4.1.2.2 Isoelectric focussing

Isoelectric focussing is a separation method in which proteins are separated in a pH gradient according to their isoelectric point, that is, the pH at which net charge of protein molecule is zero. At this pH, the protein molecules have no electrophoretic mobility. To generate pH gradient in a gel, ampholytes are used. Ampholytes get resolved on the gel according to their electric points. During electrophoresis, when proteins start migrating on the gel, they get precipitated according to their isoelectric point, where it carries a net charge zero. The proteins appear in bluish colour bands when stained with coomassie brilliant blue.

4.1.3 Agarose Gel Electrophoresis

Agarose gel electrophoresis is otherwise called as submarine or horizontal electrophoresis. This electrophoresis is used to separate DNA fragments. When DNA molecules are treated with restriction endonucleases, the various fragments that are generated can be separated from one another by this technique. Separation of DNA fragments of different lengths depends on the ability of agarose to form a molecular sieve, that allows smaller molecules to pass through more quickly than larger molecules.

4.1.3.1 Requirements

The essential requirements for submarine electrophoresis are (a) agarose; (b) buffer; (c) gel casting units; (d) separation tank; (e) transilluminator and (f) stain.

4.1.3.2 Separation technique:

Agarose gels are produced by adding an appropriate amount of agarose powder to a salt solution, and heating the mixture to dissolve the agarose. The molten agarose is pound into a casting tray which resembles a shallow, rectangular box. A comb is inserted into the molten agarose. When the gel cools and solidifies it is then immersed in buffer. The comb is then carefully removed. It leaves behind prominent wells (Fig. 4.2).

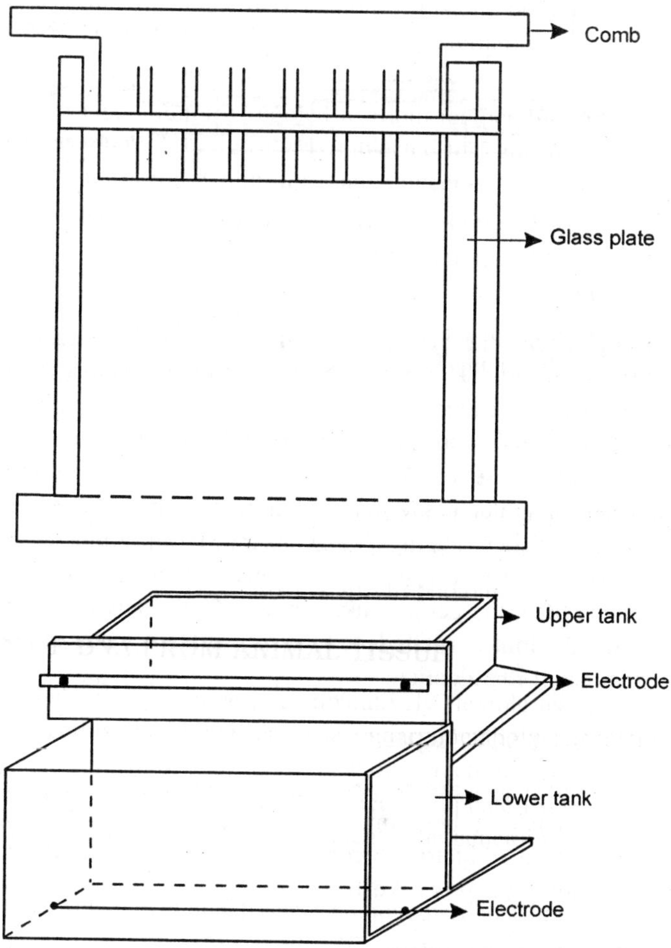

Fig. 4.1. Polyacrylamide gel electrophoresis unit.

The DNA to be electrophoresed is treated with restriction endonuclease to produce fragments of varying length. The digested DNA is mixed with sucrose and tracking dye. The sucrose increases the density of the DNA sample, so that it sinks to the bottom of the well. At least in one well a marker DNA of known length is loaded to compare with unknown fragments.

Electric current is then applied in the electrophoresis chamber. Generally current can be supplied between 50 to 150 volts. The DNA loading dye migrates out of the wells and enters the gel in response to electric field created by current. Moving of DNA into the gel can often be monitored by observing the dye front. Electrophoresis is halted when the tracking dyes move nearer to another end of gel.

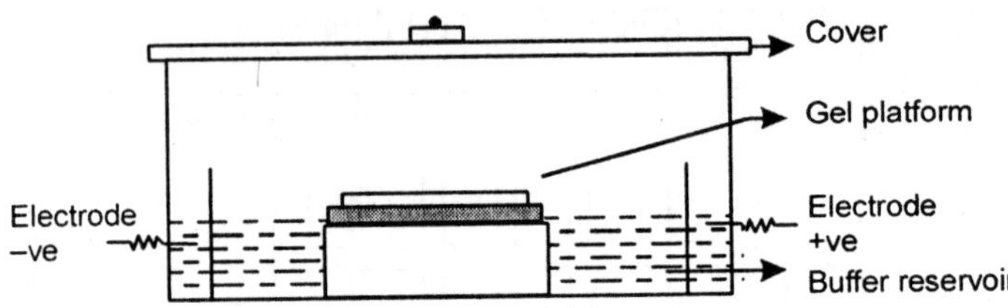

Fig. 4.2. Horizontal agarose gel electrophoresis unit.

4.1.3.3 Staining

After electrophoresis, the gel is removed and stained to make the DNA bands more easily seen. The two most commonly used dyes are ethydium bromide and methylene blue. The gel is immersed in one µg/ml solution of ethydium bromide for five to ten minutes. Ethydium bromide quickly spreads and concentrates wherever DNA molecules are present. The chemical ethydium bromide molecules actually intercalate between the bases of the DNA molecule. When the gel is exposed to ultraviolet source from transilluminator, the DNA appears as fluorescent orange bands. To stain with methylene blue, the gel is soaked in an 0.025% solution of methylene blue for 15 to 30 minutes. Methylene blue causes DNA bands to appear blue under normal room illumination.

4.2 BLOTTING TECHNIQUE:

Blotting technique is an extremely powerful tool for analyzing gene structure and used to study gene expression, once cloned cDNA is isolated. There are three important types of blotting techniques. They are: (a) Southern Blotting; (b) Northern Blotting and (c) Western Blotting.

4.2.1 (a) Southern blotting

Developed by E.M. Southern, the technique of Southern Blotting is one of the most important methods used in molecular biology. In Southern Blotting, DNA is transferred from a gel to a membrane for hybridisation analysis. In this technique, the DNA is cut with suitable restriction enzymes and run on a gel. Treatment with NaOH then denatures the DNA. The transfer of DNA from agarose gel to the membrane is performed by capillary action. The gel is placed above the buffer saturated filter paper. The nitrocellulose membrane is placed above the gel and covered by 2-3 layers of dry filter paper towel. A flow of buffer occurs through the gel and membrane to the top papers. The flow carries the DNA fragment with it. However, DNA cannot pass through the membrane and is fixed firmly to the paper. The membrane, to which the DNA is trapped, is then exposed overnight to a solution containing the radio-labelled cDNA probe. The binding of probe to its complementary sequence is then detected by autoradiography (Fig. 4.3).

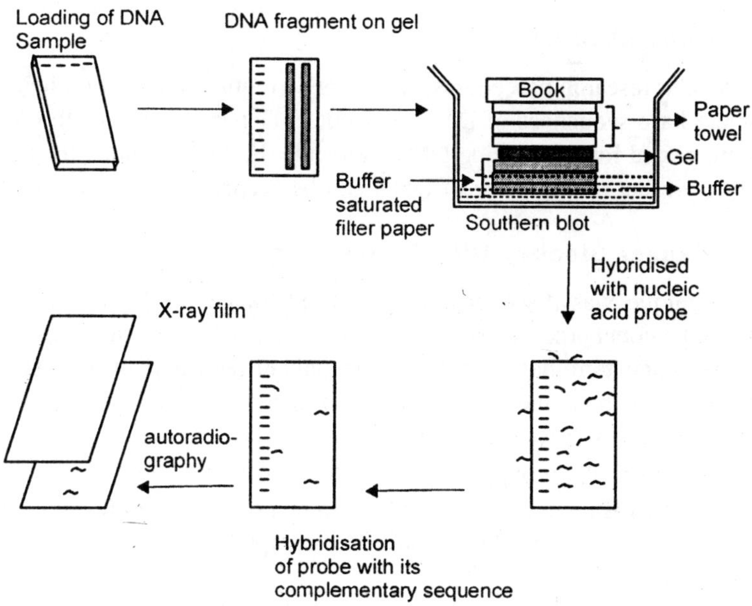

Fig. 4.3. Southern blotting technique.

4.2.1.1 Significance:

Southern blotting is useful for detecting major gene arrangements. This technique plays important role in DNA finger print, identification of novel gene, identification of structurally related genes in the species etc.

4.2.2 Northern Blotting:

Northern blotting is a technique used to analyse RNA. In northern blot, RNA is transferred from agarose gel to nitrocellulose paper for hybridisation analysis. Total cellular RNA or poly (A) RNA, is separated by size on an agarose gel. The RNA molecules in the gel are transferred to nitrocellulose or nylon membrane as described in southern blotting.

4.2.2.1 Significance:

Northern blotting is useful in the identification of a particular gene expression in a tissue or cell type. It is useful in cDNA cloning because the size of a specific mRNA can be compared with the size of cloned cDNA.

4.2.3 Western Blotting:

Western blotting is associated with the transfer of proteins from acrylamide gel to nitrocellulose or nylon membrane. This technique is useful in identification of a particular gene product. Any target protein can be identified by immunological screening method.

4.3 DNA SEQUENCING:

In molecular research procedures, scientists do require information regarding complete base sequences of DNA molecules. Generally, two methods have been suggested for sequencing DNA molecule. In this technique the number of bases present in any piece of DNA can be accurately read out.

4.3.1 Sanger Dideoxy DNA Sequencing:

This technique was discovered by Dr. Fred Sanger in 1977 and he was awarded the nobel prize. The dideoxy DNA sequence is also called enzymatic method or chain termination method. The main difference between enzymatic method and chemical method is that the former method relies on interrupted DNA synthesis, whereas the latter involves breakage of DNA molecules. The Sanger enzymatic method consists of the following steps:

(a) A single-stranded DNA fragment to be sequenced is allowed to bond with a small, radioactively labelled, complementary fragment called a primer.

(b) A preparation of primer-bound fragments is then divided into four aliquates.

(c) Each aliquate is mixed with the four nucleotides needed for synthesis of a complementary DNA strand along with one dideoxynucleotide (ddNTP), and DNA polymerase.

(d) The resulting newly synthesized DNA strand terminates wherever a ddNTP's incorporates into the sequence.

(e) After a suitable incubation time the DNA in each mixture is denatured and electrophoresed and subjected to autoradiography.

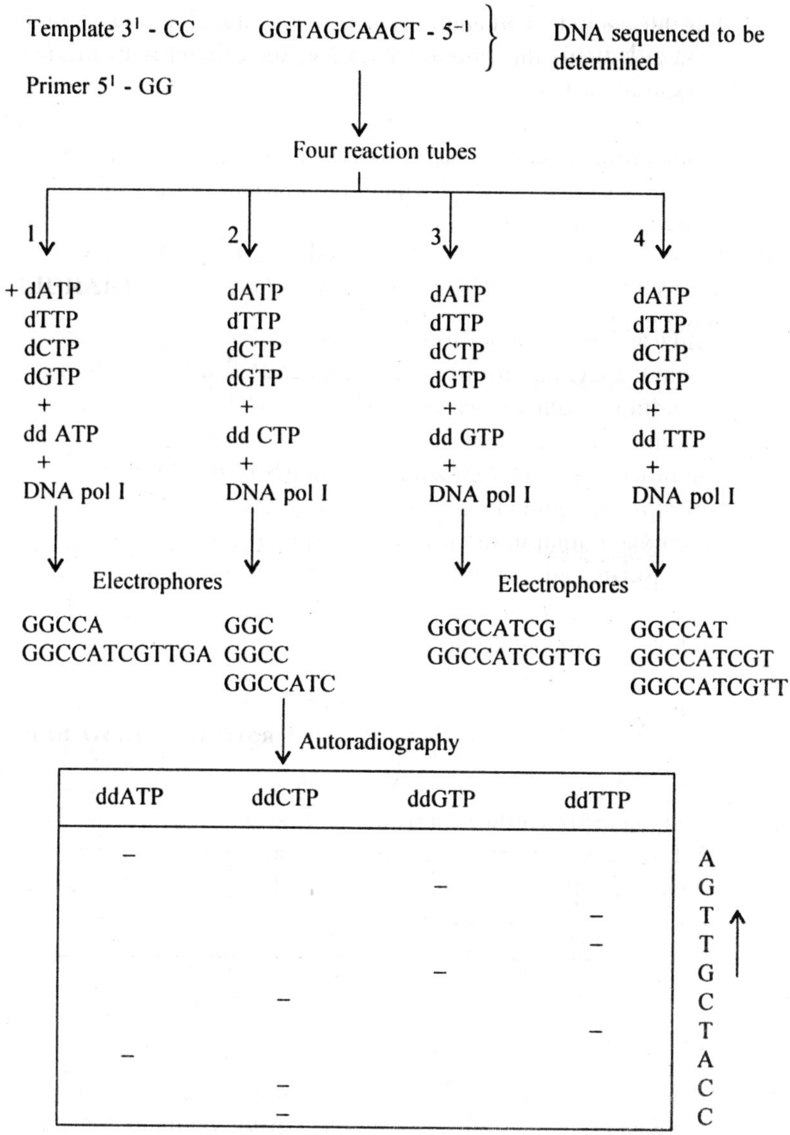

4.3.2 Maxam and Gilbert DNA Sequencing

This method is also called as chemical method of DNA sequencing and discovered by Maxam and Gilbert of Harvard University in 1977. The technique is described as follows:

(a) A DNA fragment to be sequenced are radiolabelled at one end of the molecule.

(b) After denaturation of DNA into single strand, bases are modified by chemical treatment and are divided into four reaction samples

(i) Modification of guanine base (ii) both adenine and guanine are modified (iii) the cytosine base (iv) both thymidine and cytosine bases.

(c) All the modified bases are subjected to second chemical treatment which brings breakage at the backbone of the modified base. The breakage results in the production of various length of DNA molecule.

(d) All the DNA fragments of variable length are separated on electrophoresis and DNA bands on the gel differ from one another by single base.

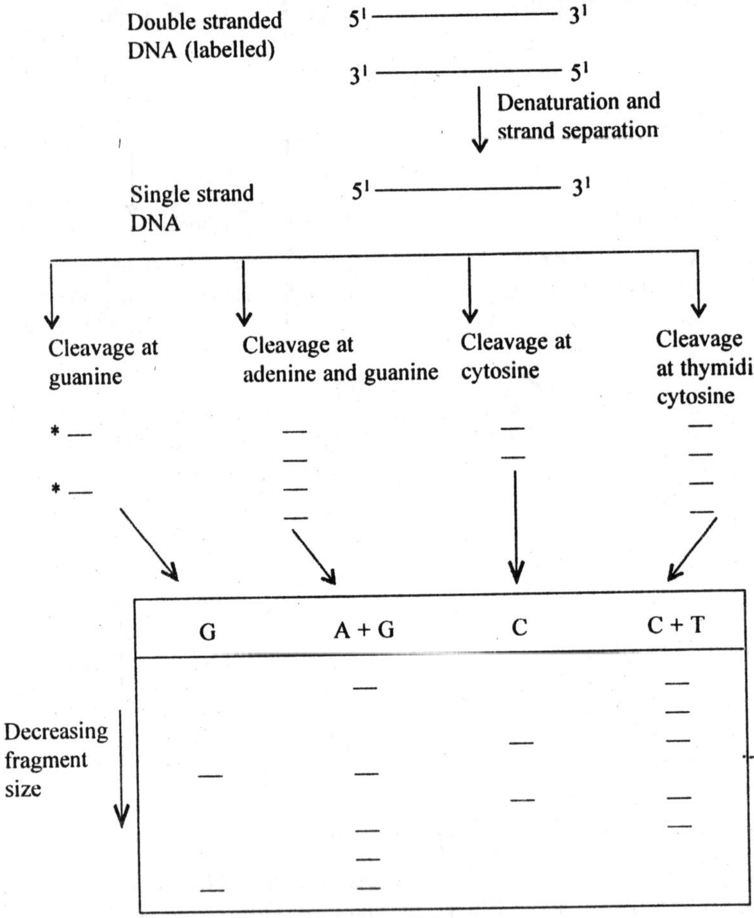

4.4 POLYMERASE CHAIN REACTION

In genetic engineering the availability of sufficient amount of DNA of interest is indispensable in order to study its genetic structure, function and manipulation. Until recently the microorganisms are the chief source to obtain multiple copies of DNA of interest. Now a new and novel cell free techniques are available to obtain millions of copies, known as polymerase chain reaction, popularly known as PCR technique. PCR is an effective technique that allows amplification of a particular DNA sequence and is able to generate innumerable copies of specific DNA sequence. The technique is so precise and powerful that even a single gene sequence can be amplified into a detectable amount. The technique was independently designed and developed by Karymullis in 1980.

4.4.1 Requirements

The essential requirements for PCR process are: (i) primers; (ii) Taq polymerase; (iii) target sequence DNA sample and (iv) four deoxyribonucleotides.

4.4.2 Technique:

PCR includes number of cycles for amplifying DNA of interest. Each cycle has three important steps:

Step 1: Denaturation: The first step is the denaturation of DNA sample in a reaction mixture to 94°C. At this thermal condition DNA strand gets separated. The temperature is maintained upto 5 mins.

Step 2: Renaturation: The temperature is lowered to 50°C to allow the two oligonucleotide primers to anneal (bind) to the complementary sequence in the DNA molecule. The time duration of 30 secs is maintained for renaturation.

Step 3: Synthesis of new DNA: The temperature is raised to 75°C. At this elevated temperature, Taq polymerase shows optimum activity and initiates DNA synthesis at 3 hydroxyl end of each primer (Fig. 4.4 and 4.5).

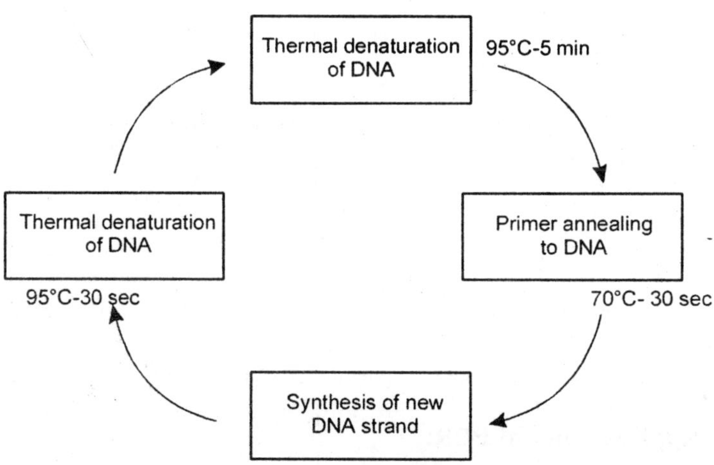

Fig. 4.4. PCR cycle.

4.4.3 Types of PCR

Several modifications of PCR have been developed for specific applications. Another reason for the modifications is to overcome problems or trouble shooters during amplification. Following are some of the improved version of PCR (i) Nested PCR; (ii) Hot start PCR; (iii) Touch down PCR and (iv) Inverse PCR:

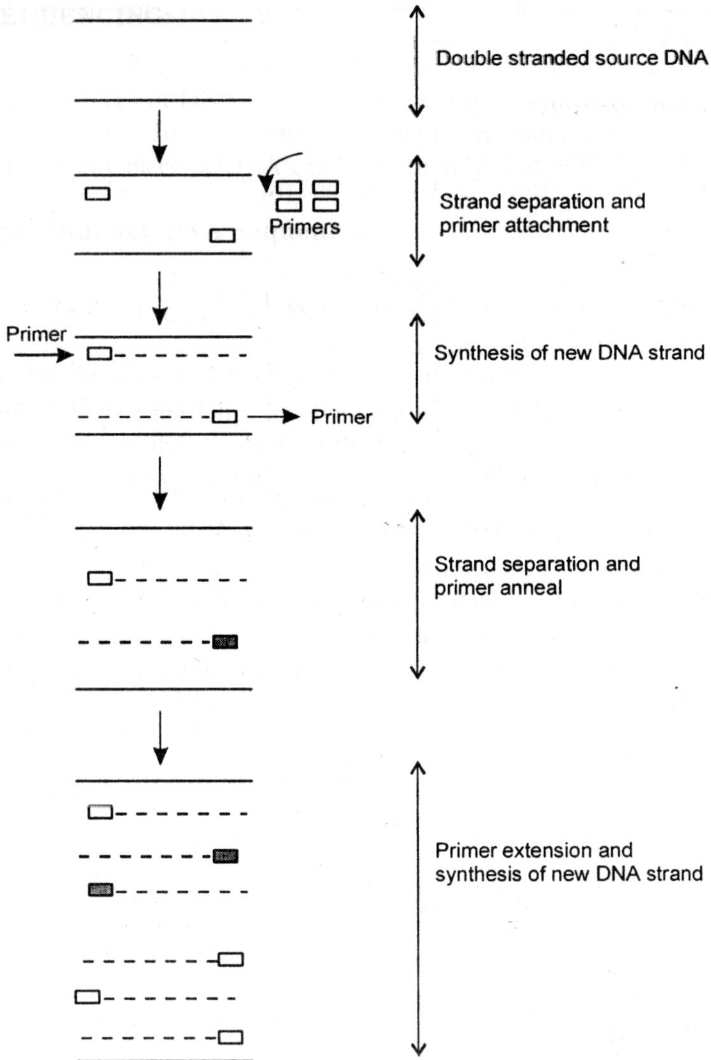

Fig. 4.5. Process of polymerase chain reaction.

4.4.4 Applications of PCR:

(a) It is extensively used for the amplification of a particular target DNA sequence.

(b) The PCR is widely employed in disease diagnosis. The speed and sensitivity of PCR helps in prenatal diagnosis of inherited disease.
(c) PCR is a basic tool for screening cloned gene in the transformation technique.
(d) In plant biotechnology PCR is used in plant disease identification, presence of any target gene in crop plants.
(e) PCR has an important role in gene manipulation and expression studies.

4.5 GENE LIBRARY

A gene library or genomic library is a collection of cloning vectors which contains genes of interest, representing an entire set of foreign DNA. Once a gene library is created, target DNA sequence present in the recombinant cells have to be identified by screening methods. In the entire gene library, host cells can be comparable to a computer hardware and inserted foreign DNA (gene) in the recombinant vector to a diskette containing stored information. During operational mode, computer takes the information, stored in diskette and output is drawn. Similarly host cells follows the instructions present in the cloning vector, and proteins are produced once a cloned gene is expressed. The gene library fulfills the goal of achievement of isolation of genes that encode proteins.

4.5.1 Types of Gene Libraries

Gene library or gene bank library is generally constructed as (i) Genomic library and (ii) cDNA library. The genetic material of prokaryotes contains continuous coding regions of structural genes. In eukaryotes, however, coding regions (exon) is separated by non coding regions (introns). Therefore different cloning strategies have to be followed during the construction of gene libraries. Two types of gene libraries are generally constructed depending on the availability of source DNA i.e., genomic DNA and cDNA.

4.5.1.1 (i) Genomic library

A genomic library represents the entire donor genome. To construct genomic library the entire DNA of an organism is cut with a restriction endonuclease, and fragments are inserted into appropriate cloning vectors. These recombinant vectors are transferred to any organism such as bacteria, virus or yeast. A target DNA sequence present in a particular cell clones are identified, subcultured and maintained as cell lines, widely known as gene banks or a clone bank.

4.5.1.2 cDNA libraries

It represents only transcribed genes, when compared to genomic libraries which contains complete set of genome. Construction cDNA library is well suited for eukaryotic structural genes.

4.5.1.2.1 cDNA preparation: During complementary DNA (cDNA) preparation poly(A) tail is used to separate mRNA from rest of tRNA and ribosomal RNA. The total RNA extraction is passed through chromatography column containing oligo dT residues, which can base pair with poly(A) tails of messenger RNA (mRNA) and retained in the column. The tRNA and rRNA molecules which lack poly(A) tail passes through the column. The mRNA is then removed from the column by adding suitable buffer. The eluted mRNA is then used to make complementary DNA. Two enzymes are required for this process.

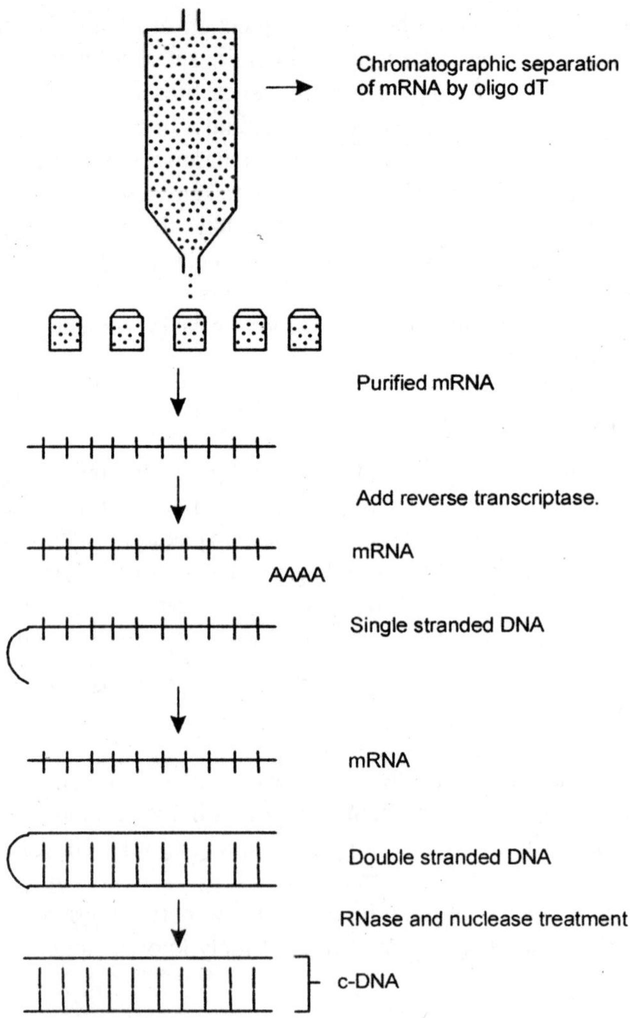

Fig. 4.6. Preparation of complementary DNA.

The purified mRNA is mixed with reaction mixture containing reverse transcriptase and four deoxyribonucleotides (dATP, dTTP, dGTP and dCTP).

Reverse transcriptase uses mRNA as a template and direct the synthesis of single stranded DNA. The synthesis of DNA in invitro is incomplete and forms a hairpin loop.

The second strand of DNA synthesis is initiated by the addition of DNA polymerase I. It first uses single stranded DNA as template and direct the synthesis of second strand. Once second strand is completed, the mRNA is hydrolysed with RNase. The hairpin loop is removed by treating DNA with nuclease enzyme and set free double stranded cDNA. The cDNA molecule is then cloned into a vector to form a cDNA library (Fig. 4.6).

4.5.1.3 Screening Gene library

4.5.1.3.1 (1) Screening by Colony Hybridisation

Principally this screening technique involves hybridisation between labelled DNA probe and a target DNA sequence. Therefore a target gene sequence can be accurately identified.

The essential steps are described as follows:

(i) The initial step in screening a library – a colony of host cells are plated onto solid medium containing antibiotics. The presence of antibiotics in the medium ensures growth of only transformed cells due to its antibiotic resistance gene in their plasmid.

(ii) When a discrete colony is formed on the plate, it is then transferred onto nitrocellulose or nylon membrane commonly referred as solid matrix. The exact position of cell colony on the plate is maintained on the matrix.

(iii) Once the cells are attached to solid matrix (nitrocellulose paper) cells are lysed, deproteinised and released DNA is denatured by alkaline reagent.

(iv) The labelled DNA probe is mixed in the next step, to facilitate hybridisation between target DNA and DNA probe. The unbound probe is washed off from the matrix (Fig. 4.7).

(v) Matrix in then subjected to autoradiography to determine the location of cells containing hybridised DNA.

(vi) Development of dark spot on the X-ray film indicates the presence of target DNA (gene). The dark spot on X-ray film corresponds to the cell colony on the master plate. Once transformed cell colony is identified, it is then subcultured and maintained as it carries cloned DNA of interest.

4.5.1.3.2 Plaque hybridisation

A genomic library is screened by infecting phage viruses containing inserted DNA on a lawn of bacteria. The steps are described as follows:

(i) A lawn of bacterial cultures are initiated on the agar plate.

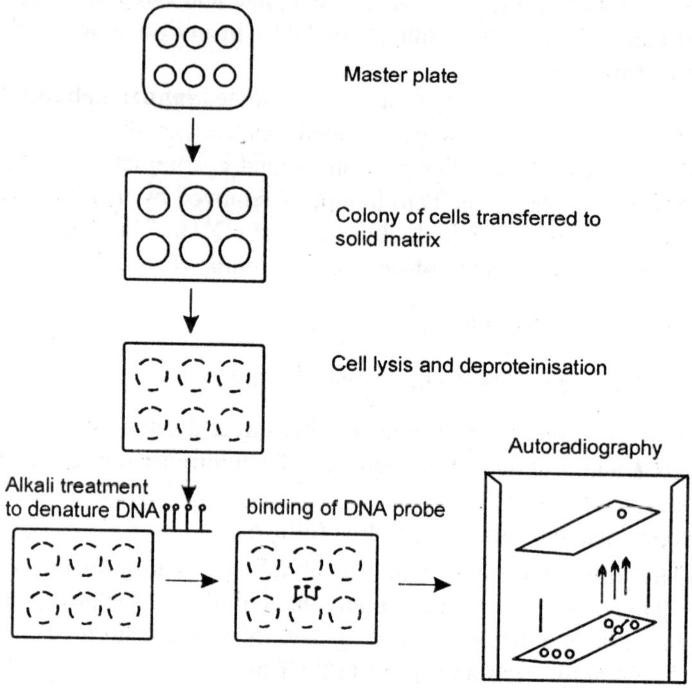

Fig. 4.7. Diagram showing colony hybridisation.

(ii) Several thousands of phage particles containing inserted DNA is spread on the lawn of bacteria.

(iii) Phage infects bacteria and plaques are obtained. These are then transferred to nitrocellulose paper, DNA is denatured to single strand and bound firmly on to solid matrix.

(iv) ^{32}P labelled DNA/RNA probe is added to ensure hybridisation.

(v) The location of the bound probe is determined by autoradiography.

(vi) The position of spot on the X-ray film corresponds to original site on agar plate. Once the desired gene is identified it is then isolated and subcultured.

4.5.1.3.3 Immunological Screening

This is another novel method of screening a gene library. This method can be used if DNA probe is not available. The cells are allowed to transcribe and translate and the presence of protein can be identified by corresponding antibody molecule. The technique is described as follows:

(i) Discrete colonies are formed on the master plate.

(ii) Transfer sample of cells from each colony to nitrocellulose or nylon membrane.

(iii) Allow the cells to transcribe and translate. The cells are then lysed and proteins are bound to matrix.

(iv) Initially primary antibody treatment is done to facilitate the binding of primary antibody to protein molecule bound on the solid matrix.

(v) Add secondary antibody that binds only to the primary antibody. Colorimetric reaction takes place only if the secondary antibody is present. (Enzyme bound secondary antibody develops coloured compound when treated with respective substrate.)

(vi) A colony on the master plate that corresponds to a positive response on the matrix is identified. The cells from the positive colony on the master plate are subcultured as this may carry insert DNA that encodes the protein that bind primary antibody (Fig. 4.8).

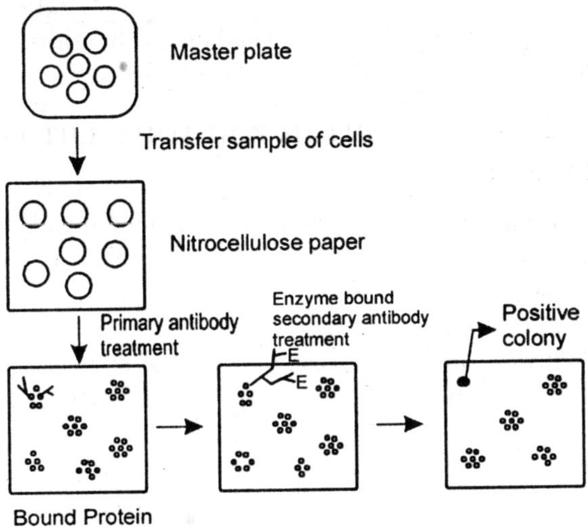

Fig. 4.8. Immunological screening.

QUESTIONS

1. Describe the techniques of Gel Electrophoresis.
(BU April 2001)
2. Describe the techniques of Agarose gel electrophoresis.
3. Explain the role of electrophoresis in genetic engineering.
4. Add a note on the techniques of polyacrylamide gel electrophoresis.
5. Give an account of PCR and its applications.
(BU Nov. 1997)
6. Give a brief account of PCR. (May 1996)
7. Give an account on polymerase chain reaction.
(BU Nov. 1999)
8. Briefly describe the technique of "Southern blotting". Add a note on its applications.
(BU Nov. 1997)
(BU April 2000)
(BU April 2002)

9. Give an account of PCR, describing the reactions and applications. (BU Nov. 1999)
10. Give an account of Gene Library. Add a note on its significance in genetic engineering.
11. Explain how cDNA is prepared. Add a note on its screening methods. (BU April 1998)
12. Give an account of cDNA library. (BU April 2000)
13. Describe various methods of screening transformants. (BU May 1996) (BU Nov. 1999)
14. Explain briefly colony hybridisation and plaque hybridisation.
15. Explain Sanger Method of DNA Sequencing. (BU April 1998)

SHORT QUESTIONS

1. Southern blotting
2. Autoradiography
3. cDNA Library
4. Colony hybridization

5

Introduction of DNA into Living Cells

The area of gene transfer into living organism has seen exciting progress. The vectorless method of introducing DNA into living organisms gained much momentum. The use of biological vectors to transport gene of interest exhibit several constraints. Agrobacterium mediated gene transfer is most efficient only to dicotyledons. Most cereals, including legumes are highly recalcitrant to agrobacterium method of gene transfer. In addition, plant cell wall is an efficient barrier and traps DNA molecules. So far only viruses are used as vectors to transfer gene in animal system. However, the risks of contamination and mutation in viral DNA cannot be ruled out. Hence, several direct methods of gene transfer system have been designed.

5.1 MICROINJECTION

Microinjection is a novel and efficient technique in which DNA is delivered directly inside the living cell. Microinjection uses microcapillaries and microscopic devices to deliver DNA into cells in such a way that the injected cell survives the treatment and is able to proliferate in culture. Two types of pipettes are used, one is micro pipette to deliver DNA and another is holding pipette. The exogenous DNA of 1 μm is taken in micropipette or microinjector, the cells or protoplast are firmly immobilised by holding pipette which exerts suction pressure (Fig. 5.1). Nearly 10-15 cells can be injected per minute. Several reports claim to have achieved more than 50% transformation by microinjection. The whole operation can be visualised on screen *by fixing* video camera to microscope.

Advantages

(i) The quantity of DNA delivery can be optimized.

(ii) DNA delivery is predictable, even into the cell nucleus and
(iii) Even a smallest cell can be targeted to deliver DNA.

Disadvantages

(i) Requires skilled persons for microinjection.
(ii) Only one cell receives DNA per injection.
(iii) Costly instruments.

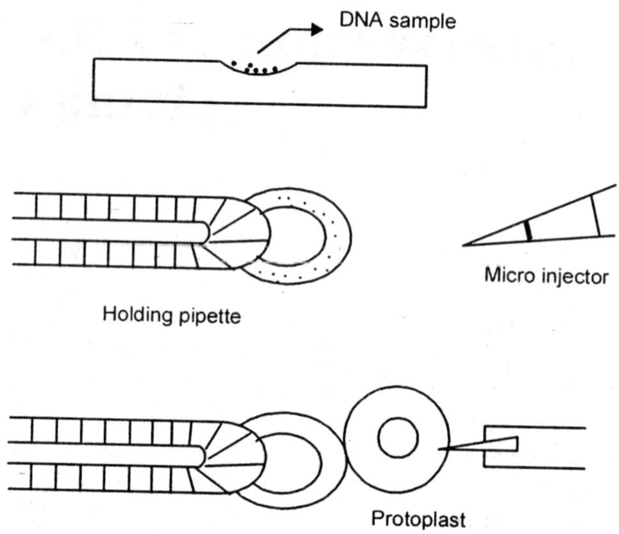

Fig. 5.1. DNA introduction by Microinjection.

5.2 ELECTROPORATION

Electroporation facilitates cells to uptake DNA molecules by reversibly altering the permeability of cell membrane. Electroporation is one of the several standard techniques for efficient transformation. The technique is more suitable for the transformation of plant protoplast and animal cells. A recombinant plasmid is mixed along protoplast in a container. A current of field strength 300 to 800 volts/cm is given to the protoplast suspension. The transient holes are formed on the protoplast following electroporation (Fig. 5.2).

Advantages

(i) Efficient transformation.
(ii) Obtain large number of transformed cells.
(iii) May not require skilled persons.

Disadvantages

(i) Requires protoplast for transformation.

(ii) Difficulties associated with protoplast regeneration.
(iii) Risk of obtaining genetic variation in protoplast regenerated plants.

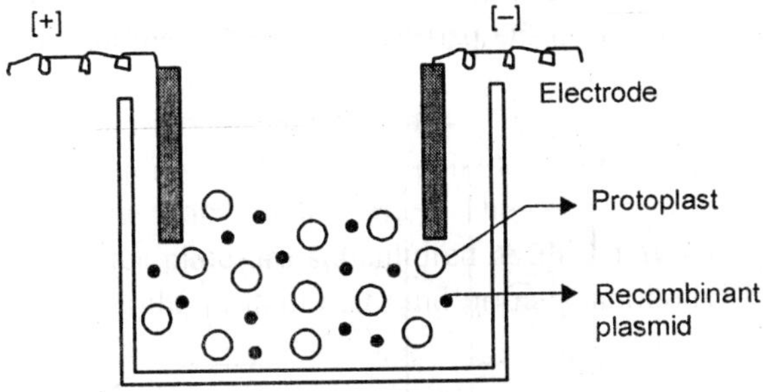

Fig. 5.2. Electroporation.

5.3 MICROPROJECTILE SHOT GUN METHOD

Shot gun or Microprojectile bombardment method is prominently used for plant transformation. The problem associated with protoplast regeneration is well known. Hence, biolystic gun method is the most advanced technique used in gene cloning. Any plant tissues, organs, cells and even pollen can be transformed very efficiently.

This technique consists of basically two types of projectile systems, macro and microprojectile. The DNA sequence of interest is precipitated and coated on micro metals such as tungsten or gold particles of size 1-3μm. This is called as microprojectile. These DNA coated micrometals are then placed before a plastic or bullet known as macroprojectile.

Inside the gene gun barrel both the projectile systems are placed and with the help of a firing device in a vacuum condition macroprojectiles are fired against retained plate. High pressurised nitrogen is applied to drive the macroprojectile which alongwith the microprojectile hits the retaining plate. The retaining plate allows microprojectile to move and retains the macro-projectile. Under high explosive discharge the high velocity microprojectile hits the tissues or cells placed next to it on the stage. Once microprojectile enters the cells, the DNA gets dissociated from the micrometals (Tungsten or gold particles) and integrate into host DNA. The cells present only in the line with the microprojectile movement gets damaged. However, rest of the cells survive and are transformed efficiently (Fig. 5.3).

Advantages

(i) Direct delivery of DNA in the nucleus.
(ii) One shot can lead to multiple hits i.e., transfer of genes into many cells.
(iii) Speedy and accurate result.
(iv) Large number of DNA molecules can be introduced.

(v) Transformed plant organs (shoot tip, embryos) display efficient regeneration.
(vi) No risk of genetic variation.
(vii) The genes coated on the particle have biological activity.

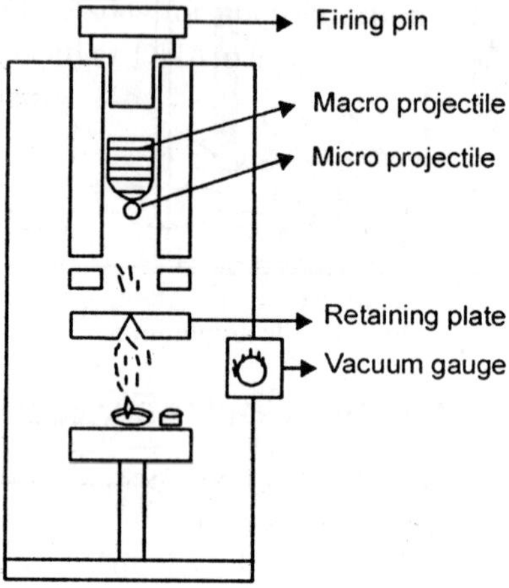

Fig. 5.3. Showing gene transfer by gene gun method.

5.4 ULTRA SONICATION

It is a simple method facilitating DNA uptake by directing ultrasound on the target cells. Small pieces of tissues or cells and known amount of exogenous DNA are taken in a vial, containing appropriate medium.

The microtip of sonicator is immersed in the suspension and pulses of ultrasound is delivered by sonicator. The vigorous vibration in the medium and violent collapse of bubbles generate high pressure and shock waves may result in sporadic localised rupture in the membrane and it can uptake exogenous DNA. This method has been successfully used in the transformation of several plants.

5.5 LIPOSOME FUSION

Fusion of DNA-containing liposome have been applied to various tissues and cell cultures. In this technique it has been shown that liposomes can carry small dye molecules into cells within tissue and presuming that it can carry genes into the target cells, "DNA or RNA encapsulated in liposomes, can be transferred into plant protoplasm by direct fusion with plasma membrane". As there is not much advantage with this method, it is not much in use.

5.6 MICROLASER

The basic principle lying in this technique is localised rupture of cell membrane. A microlaser beam is focussed into the light path of a microscope and it can be used to burn holes into cell walls and membranes. The localised rupture or burn hole region permits the entry of exogenous DNA. The extent of damage on the membrane is reversible.

Table 5.1. Comparison of DNA delivery methods

Methods		Comment
1.	Ti plasmid mediated	Highly effective, limited to dicotyledons
2.	Microprojectile gene gun	Used with wide range of plant tissues, efficient in yielding a stable transformation
3.	Viral vector	Effective for animal system, except for few examples, not an effective method for plant system
4.	Micro injection	Efficient for plant and animal systems requires skilled workers
5.	Electroporation	Its use is generally limited to plant cell protoplast.
6.	Liposome fusion	Can be used only with plant cell protoplast

5.7 THE UPTAKE OF DNA BY BACTERIAL CELLS

Most species of bacteria are able to take up DNA molecules from the medium in which they grow. A DNA molecule taken up in this way will degrade, but occasionally it may survive and replicate in the host cell. Most species of bacteria, including *E. coli* takes up only limited amount of DNA under normal circumstances. In order to increase the uptake efficiency, the bacteia will have to undergo some form of physical and chemical treatments that enhance their ability to take up DNA. Cells that have undergone this treatment are said to be competant.

5.7.1 Preparation oc Competant *E. coli* Cells

It was observed that *E. coli* cells that had been soaked in ice-cold salt solution were more efficient at DNA uptake than unsoaked cells. A solution of 50mm calcium chloride is traditionally used. Calcium chloride causes the DNA to precipitate on to the outside of the cells or some kind of change in the cell wall that improves DNA binding. The actual movement of DNA into competant cells is stimualted by raising the temperature to 42°C.

Selection is based on the use of selectable marker carried by the plasmid. A good example is the ampicillin resistance gene of pBR 322. After transformation experiment with pBR 322, only those *E. coli* cells that have taken up a plasmid will be $amp^R.tet^R$ and are able to form colonies on agar medium that contains ampicillin or tetracycline, whereas non-transformants, which are still amp^S tet^S, will not produce colonies on the selective medium.

5.8 INTRODUCTION OF PHAGE DNA INTO BACTERIAL CELLS

5.8.1 Invitro Packaging

Recombinant λ molecules could be packaged into λ head and tail structures in the test-tube. Packaging requires number of proteins coded by λ genome. These can be prepared at a high concentration from cells infected with defective λ phage strains. The defective λ phage carries mutations so the defective phage cannot replicate, though it does direct synthesis of all the proteins needed for packaging. The proteins accumualte in the bacterium and can be purified from cultures of *E. coli* infected with mutated λ, and are used for in vitro packaging of recombinant λ molecules.

QUESTIONS

1. Describe various methods of introducing DNA into living cells.
2. Explain different physical methods of gene transfer.
3. What is gene cloning? Give an account on various gene transfer methods.

   (BU Nov. 1999)
4. Write an account on microinjection and Biolystic gene gun method of gene transfer.
5. Elucidate various methods of introducing DNA into living cells.

   (BU April, 2001)

6

Genetic Engineering of Plants

Imagine! Infant eating banana which was engineered to produce vaccines for infectious disease. The antibodies can be harvested from potato plants! Plants can exhibit defensive strategy by marshalling toxin protein against pathogenic pests. Increase in shelf life of fruits and vegetables are some of the present and future realities in genetically engineered plants or Transgenic plants. The term "transgenic" refers to a plant containing foreign gene(s). Transgenic plants have definite advantage over hybrids produced by conventional methods. Transgenic plants play a multirole in improving agronomic traits (Table 1).

Table 6.1. Transgenic crops approved in USA for commercial use

Product	Genetically altered traits	Company	Year of approval and product name
Tomato	Delayed ripening	Calgene	1994, Flavar Savr
	Delayed ripening	DNA plant	1995, Endless summer
Summer	Delayed ripening altered pectin content	Tech Monsanto	1995
		Zeneca petoseed	1995
Cotton	Bt gene incorporated plants	Monsanto	1995, Bollgard
	Resistant to glyphosate	Calgene	1996, Round up
	Resistant to Sulphomoyl urea	Dupont	1996
Soybean	Resistant to glyphosate	Monsanto	1995, Round up Ready
Maize	Bt gene incorporated	Ciba-Geigy	1996
	Resistant to glufosinate	Devalb, Agro Evo	1996, Liberty Link
	Male Sterility	Plant Genetic Sys	1996
	Bt gene incorporated	Novartis	1996 Yield Gard
Rapeseed Canola	Altered oil composition (high lauric acid)	Calgene	1995, Laurical
Squash	Resistant to virus	Asgrow	1995, Freedom

6.1 METHODS IN PLANT GENETIC ENGINEERING

Agrobacterium Ti Plasmid

The feasibility of *Agrobacterium tumeifaciens* as a cloning vector in engineering of plants is remarkable.

Agrobacterium tumeifaciens is a gram negative soil bacterium, popularly called as "Natural Genetic Engineer". In nature, *Agrobacterium tumeifaciens* infects dicotyledons and causes a disease known as crown gall.

Most of the *Agrobacterium* strains possess a plasmid known as Tumor-inducing plasmid (Ti plasmid) of 200 kb in size. Crown gall formation is the direct result of transfer of portion of Ti plasmid into the plants. The general features of Ti plasmid are as follows:

(i) Ti plasmid contains unique set of genes responsible for transformation of plants.

(ii) The transferred portion of Ti plasmid is referred as T-DNA (Transfer DNA).

(iii) The vir genes are located on 40 kb of the Ti plasmid that lies outside the T-DNA. The product of vir genes are responsible for T-DNA transfer.

(iv) T-DNA contains unique set of genes responsible for the production of plant hormones. Once a successful integration takes place, expressed genes are responsible for tumour production.

(v) Ti plasmid also possesses genes for opine catabolism and derive energy from it.

(vi) The left and right border of T-DNA exhibits 25 kb unique sequence of direct repeat and are invariably transferred to plant cells (Fig. 6.1).

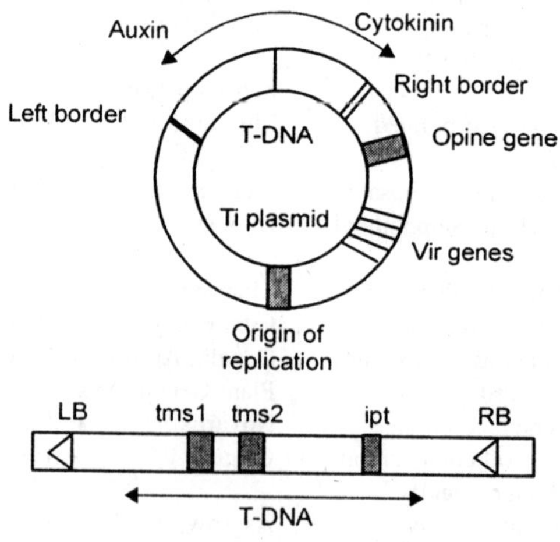

Fig. 6.1. Ti plasmid.

6.1.1 Mechanism of T-DNA transfer

One of the most fascinating process that occurs in nature is the mechanism of T-DNA transfer from bacterium into the plant cell. The role played by vir genes in transferring T-DNA into plant cell is exemplary. A group of seven to eight vir genes are organised as operons. Vir genes are classified as virA to virH. During infection, in response to signal molecule produced by plant system, all the vir genes are expressed and play guided role in removing T-DNA region from Ti plasmid. More precisely, the right border of T-DNA is recognised by the virgene product (virD and C).

These proteins cut the sequence accurately at right border releases single stranded T-DNA. The T-DNA strand on either side is escorted by virE proteins as T-complex. D_2 proteins occupy if the front position of T-complex and successfully transfer from bacterial cell into plant cell. Once this enters cytoplasm, integration takes place inside the host DNA (Figs. 6.2 and 6.3).

Construction of *Agrobacterium* cloning vector

Ti plasmid contains disease causing genes in T-DNA region. Hence, it is very essential to remove T-DNA portion, before splicing a gene of interest. However, right and left border sequence in T-DNA is retained as it plays a significant role in guiding T-DNA-transfer.

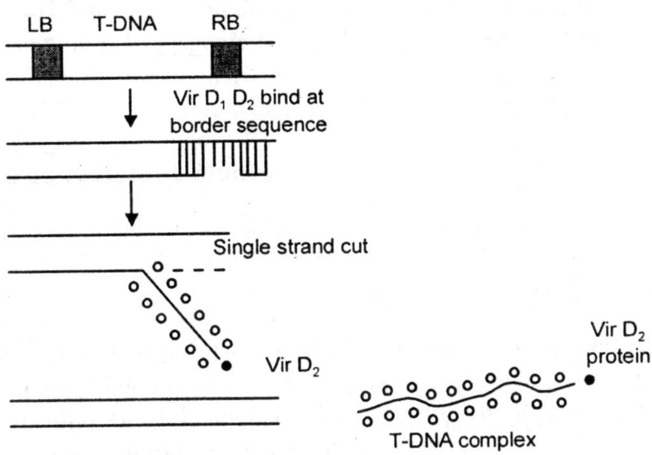

Fig. 6.2. Release of T-DNA.

Two approaches are used to modify *Agrobacterium* Ti plasmid into a cloning vector.

6.1.2 Binary vector

The cloning vector is designed in such a way that it contains both origin of replication of *E. coli* and *A. tumeifaciens*. T-DNA is removed after retaining

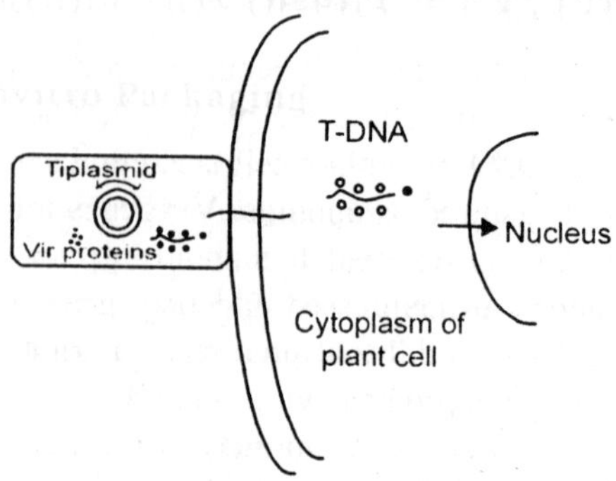

Fig. 6.3. T-DNA transfer from bacterium to plant cell.

right and left border sequence. The DNA of interest (foreign gene) will be inserted between left and right border of T-DNA. This binary vector is then mobilised from *E. coli* to *A. tumeifaciens. Agrobacterium* is also modified as defective plasmid, i.e., T-DNA is removed but vir genes are retained. The vir gene in the defective plasmid synthesises proteins that involve in transferring T-DNA from bacteria to plant cell chromosome (Fig. 6.4).

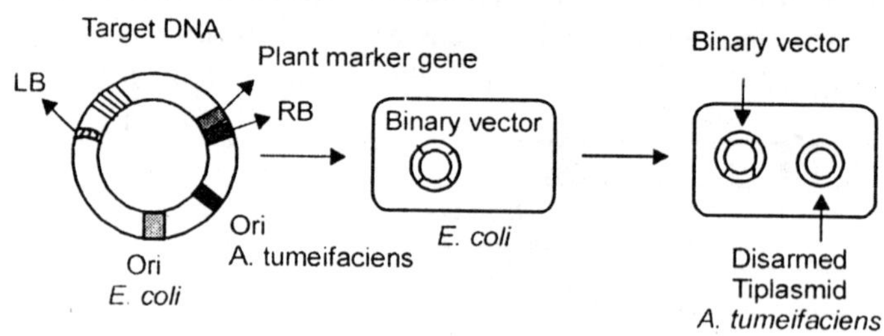

Fig. 6.4. Binary vector method of gene transfer.

6.1.3 Co-integrative vector:

A typical cointegrative vector has the following sequence. The cointegration vector carries only an origin of replication of *E. coli*, plant selectable markers such as NPT II gene, homologous sequence and T-DNA right border and a target gene. The cointegrative vector containing foreign gene is mobilised into *A. tumeifaciens* by triparental mating. *A. tumeifaciens* contains disarmed Ti plasmid retaining left border vir genes and homologous sequence. The presence of homologous sequence on both the plasmid shares a common site for homologous recombination. Following recombination, cointegrative vector becomes part of the defective plasmid. Vir genes then mobilise T-DNA (containing foreign gene) from *Agrobacterium* into plant cell (Fig. 6.5).

Plant cell transformation, especially in cereals is carried out by electroporation, biolystic gun and micro injection which have been described in the earlier chapter.

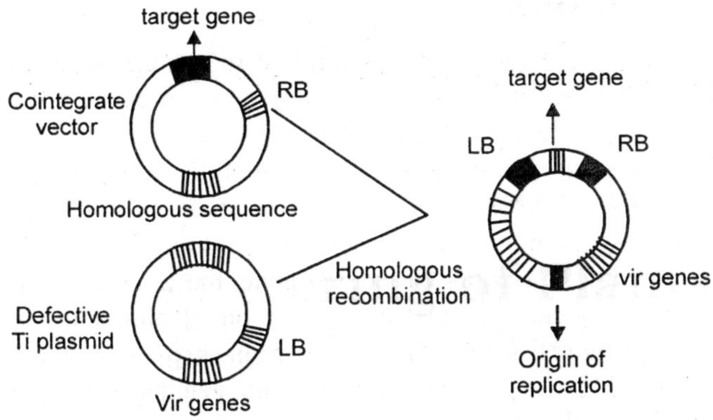

Fig. 6.5. Cointegrative method of gene transfer.

6.2 APPLICATION OF TRANSGENIC PI ANTS

6.2.1 Insect-resistant plants

Insects are harmful to crop productivity. Extensive use of chemical insecticides are hazardous. Plant genetic engineering approach for insect resistance is specific. The insecticidal toxin gene is introduced into the plants and the expression of cloned gene produces a toxin, which is lethal to insects. The classical example is Transgenic Bt cotton, in which an insecticidal Bt toxin gene from *Bacillus thuringiensis* have been successfully cloned. Efficient expression of Bt crystal protein can control insects by paralysing its midgut.

6.2.2 Virus-resistant plants

Viruses can cause substantial crop loss. The use of chemical spray cannot control completely. The meristem culture has been tried for controlling few viruses.

Genetic engineering strategy for virus resistant plants have effectively controlled several viruses. The cloning of Tobacco Mosaic Virus (TMV) coat protein gene in transgenic tobacco plant showed high level of resistance to TMV infection. In RNA mediated protection only infected plant cells are killed by degrading all RNA molecules by specific expression of ribonucleases cytotoxic enzyme.

6.2.3 In post-harvest technology

Senescence in fruits and flowers induce ageing process. Softening and ripening are the two physiological changes associated with spoilage in fruits and vegetables. By using antisense RNA technology, ripening and softening process can be delayed by blocking the synthesis of ripening related enzyme (Acc synthase and α polygalacturonase). The calgene company marketed transgenic tomatoes with the trade name "flavrsavr".

6.2.4 Pharma Plants

Pharma plants are Transgenic plants tailored to synthesise antibodies and vaccines. If plants are successful as green factories of edible vaccines, then the vaccination problem of Third World countries can be solved. Edible vaccines do not require special form of transportation, refrigeration or needles for administration. Edible vaccines may be used to protect against the diarrhoea-causing pathogen *E. coli.* Upon consumption of transformed potatoes containing *E. coli* protein coding gene, the *E. coli* proteins formed in the human body then induce the production of antibodies against the proteins, conferring resistance. Apart from vaccines, plants can also be made to produce an edible form of the hormone insulin. Feeding transgenic potatoes containing the hybrid human insulin genes cholera toxin subunit gene to mice delayed the onset of symptoms by few weeks.

Mass production of expensive monoclonal antibodies in plants is also possible. These antibodies can be used to treat various diseases. Antibodies made in soya plants prevent infection of mice by herpes virus and antibodies made in tobacco helps to prevent the bacterial infections. Antibodies produced from plants are purer, safer and less expensive compared to those made from animals.

6.2.5 Biopolymer Production

As plastics are currently synthesized from non-renewable sources such as petroleum products, researchers are constantly in search of alternative methods. One particular approach involving the insertion of three poly-3-hydroxy-butyrate (PHB) – related genes into *Arabidopsis* has not been successful due to its brittle nature. In order to overcome this problem four different genes have been transferred into *Arabidopsis* to produce a copolymer of PHB and poly-3 hydroxy valerate (PHBV).

QUESTIONS

1. Explain the role of *Agrobacterium tumeifaciens* in plant genetic engineering.
2. Describe vector mediated gene transfer in plants.

   (BU Nov. 1999)
3. Discuss *Agrobacterium tumeifaciens* as cloning vector.

   (BU April 2001)
4. Discuss different applications of transgenic plants.
5. Compare and contrast Binary vector from cointegrative vector in plant transformation.

7

Genetic Engineering of Animals

The livestock improvement and commercial exploitation of genetically altered animals for the production of useful recombinant products are some of the major goals in biotechnology. Although conventional animal breeding methods have been in practice, for better traits such as growth rate, milk production, superior wool quality etc., genetic engineering offers a novel technique of introducing useful gene into fertilised egg. The presence of foreign gene in animals are called as transgenic animals.

The idea of molecular pharming have now become attractive in the field for the production of recombinant product in milk. Several transgenic farm animals such as cow, sheep, cattle etc. can be effectively exploited for the production of therapeutic proteins secreted into the milk in mammary gland. Genetic manipulation of animals offers significant contribution in understanding mammalian gene expression, combating diseases, gene therapy etc.

7.1 GENE TRANSFER METHODS

Although transgenic mice were produced ten years ago, the mouse is still the animal of choice and used as model system for most of the experiments. However, currently pigs are used for most experiments because of its potential transgenic frequency. The general cloning strategy for animals are as follows:

The introduction of a novel gene into the nucleus of a fertilised egg. The fertilised egg carrying transgenes are implanted into receptive female (foster mother). Implanted females or foster mother gives birth to progenies carrying transgene. Finally the animals which carry transgenes are identified and bred to establish transgenic lines. Following are some of the gene transfer methods carried out in animals.

7.1.1 Retroviral method

In retroviral method, eight cell stage embryo is removed and infected by defective retrovirus carrying foreign gene. The retrovirus is made defective in such a way that, it can effectively penetrate cells without replication. The transformed embryo is then implanted back in foster mother. After birth, the matings are carried out to establish transgenic line.

Although retrovirus is a good choice as a vector, it can carry only a limited size of the foreign DNA.

7.1.2 Microinjection

In another approach fertilised eggs are microinjected by DNA. Microinjection requires excess of fertilized eggs. To obtain good number of eggs, donor female is induced to superovulate and fertilised eggs are obtained after mated with male (Fig. 7.1).

Transgene is microinjected into the male pronucleus of the fertilised eggs. Male pronucleus is clearly visible as it is larger than the female pronucleus in the fertilised egg. Later, these eggs are microsurgically implanted into the foster mother cell. The transgenic lines are established from transgenic pups.

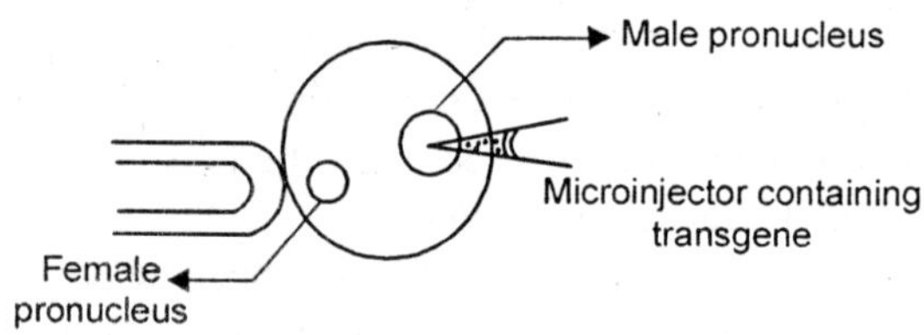

Fig. 7.1. Microinjection.

7.1.3 Embryonic stem cells (ES)

These are pleuripotent cells of early embryo (blastocyst) capable of giving rise to all differentiated cells, including germ line cells. Embryonic stem cells (ES) can be cultured, maintained and are able to proliferate. The method of producing transgenic mice by genetically modified embryonic stem cells are as follows:

(i) Embryonic stem cell (ES) culture is initiated.
(ii) The ES are transformed and screening is done in PCR analysis.
(iii) Transformed cells can be cultured and inserted into blastocyst, which are then implanted in foster mother.
(iv) Transgenic lines are determined by crosses from founder mice.

7.2 APPLICATIONS OF TRANSGENIC ANIMALS

7.2.1 Transgenic animal pharming:

Several dairy animals like cattle, sheep, goat and buffaloes can be used as

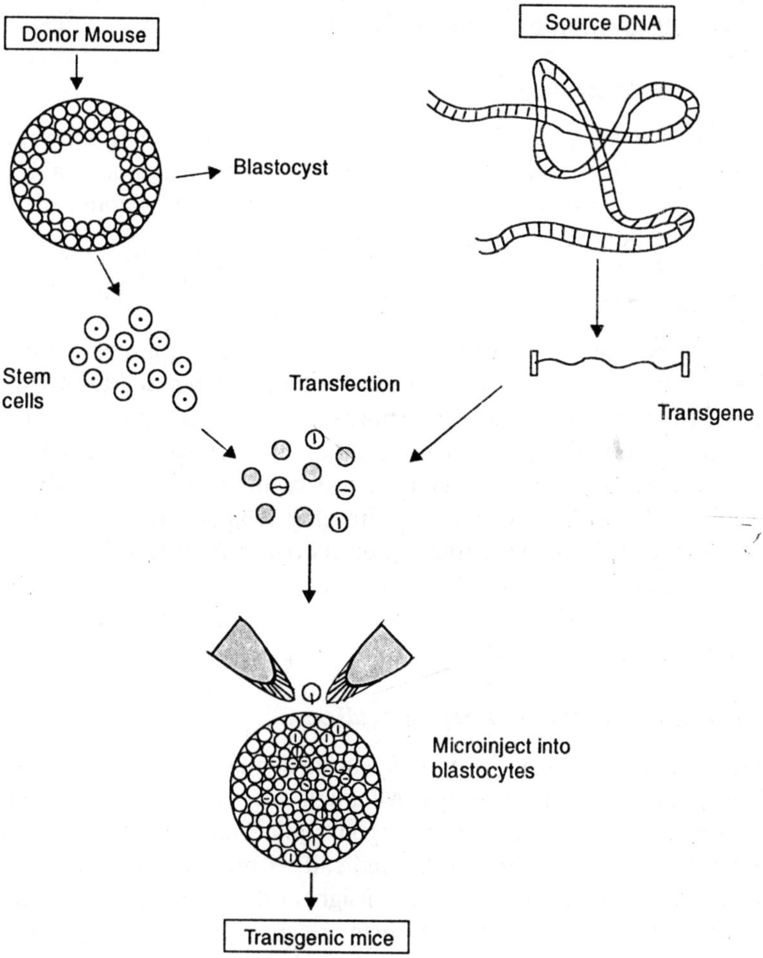

Fig. 7.2. Embryonic stem cells method of gene transfer.

bioreactor to produce recombinant proteins. The dairy cattle produces several thousands of litres of milk per year. The milk contains proteins like casein and β-lactoglobulin. Production of these proteins is tightly regulated by promoters that limit gene expression only to mammary glands. The transgene for therapeutic proteins are tagged with this mammary gland specific promoter and the expressed proteins are then secreted into milk. Over expression of casein transgene results in over production of casein protein which is feasible in cheese production.

Livestocks are prone to various diseases, bacterial disease such as mastitis in cattle and fowl cholera in some breeds. Although diseases are controlled by vaccination, genetic engineering offers novel approach to introduce and express transgene for monoclonal antibodies as a immune protection. Transgene for interleukin clotting factor 1X, cystic fibrosis transmembrane regulator (CFTR) have been cloned and expressed under the control of mammary gland specific promoters.

7.2.2 Transgenic pigs

Pigs are selectively chosen to introduce transgene, due to its shorter gestation period. The pigs carry transgene for human growth hormone (hGH) and β-globulin gene for human haemoglobin. Once these were expressed, accumulated in blood and later purified.

7.2.3 Transgenic fish

Transgenic technology in fish is concerned with studying growth hormones. In fishes, transgenes can be introduced into fertilised egg and further development of embryo takes place in temperature controlled environment. Further, fertilisation in fishes is external and hence implantation into the female is not required as in case of mammalian system. Some fishes like winter flounder exhibit unique tolerance to freezing temperature due to its antifreeze gene. The antifreeze protein protects fishes from freezing under extreme cold conditions in the sea. In one study, antifreeze gene was isolated and successfully cloned in salmon fishes for freeze tolerance, and gene for disease resistance will be introduced into temperate and tropical fishes.

7.2.4 Experimental Transgenic Mice

Transgenic mice is one of the most favourite animal for the investigation of various diseases of human beings. Although mice is a mammalian system, the information procured from transgenic mice may not be accepted in medical field or in other words medically irrelevant. But mice as an experimental animal can throw critical insight in studying etiology of diseases. Several genetic diseases such as Alzheimer's disease, arthritis, hypertension and muscular dystrophy have been investigated in transgenic mice as a model system.

In one of the classical investigations researchers have developed transgenic mice for Alzheimer's disease. It is a degenerative brain disorder characterised by gradual loss of memory and thinking ability, accompanied by personality change, reducing physical activity etc. During etiology of the disease, it was found that, 4- k dalton protein called βA4 or β protein, a principal component of senile plagues developed, gets accumulated at the ends of neurites and thus resulted in loss of brain cells. Further investigation reveals that some strains of mice produce senile plagues during their life span, whereas others do not. Thus transgenic mice is important for studying a transgene encoding β-protein, which might provide a model for studying the molecular basis of Alzheimer's disease.

In one more approach, transgenic mice was produced to express and secrete cystic fibrosis transmembrane regulator (CFTR) protein into milk.

Substantial quantities of CFTR protein is required to study its function and for therapy in treating cystic fibrosis. It is a genetic disease in which defective cystic fibrosis gene produces altered protein that may fail to regulate the flow of chloride ion, as a result, thick mucus gets accumulated, especially in lungs. The mucus could be a convenient site for bacterial infection. Therefore production of CFTR protein in model transgenic mice not only facilitates the understanding of this protein but also used as potential therapeutic agent.

7.2.5 Transgenic Lamb

UK based institute animal physiology and genetic research has produced transgenic lamb for cloned protein production. The gene for alpha-antitrypsin was successfully cloned in the lamb. (In humans, deficiency of alpha-antitrypsin (αAT) inhibits the enzyme elastase that digests foreign particles and facilitate in clearing of lungs. In absence of αAT, elastase enzyme is not inhibited and therefore diggest lung tissue). Hence production of αAT in transgenic lamb could be effectively used as therapeutic agent.

Table 7.1. Mammary gland transgenes, promoter sequences and recipient organisms

Transgenes	Promoter	Transgenic animals
α-Antitrypsin	β-Lactoglobulin	Sheep
Tissue plasminogen activator	whey acidic	Goat
Clotting factor IX	β-Lactoglobulin	Sheep
Lacto ferin	αS1-casein	Cattle
Urokinase	αS1-casein	Mouse
CFTR	βS1-casein	Mouse
Interleukin	β-casein	Rabbit

QUESTIONS

1. Explain the role of genetic engineering for livestock improvement.
2. Explain the significance of transgenic animals.
3. Describe various methods of producing transgenic animals.
4. Discuss various applications of transgenic animals.

8

Applications of Cloning in Gene Analysis

8.1 HOW TO OBTAIN A CLONE OF A SPECIFIC GENE

In recombinant DNA technology, obtaining a clone of a specific gene is a challenging task. Construction of recombinant DNA molecule and transformation of living cells can be used to obtain specific cloned genes. A small bacterial cell *E. coli* contains several thousand genes and a restriction digestion of total cell DNA will produce not only the fragments carrying a specific gene, but also many other fragments carrying other genes. Since, there is no selection method at ligation stage, several recombinant molecules are produced. Hence, variety of recombinant clones are obtained after transformation. Therefore suitable molecular strategy have been enforced to obtain a clone of a specific gene.

Currently, there are two basic approaches designed to obtain a clone of interest.

A. Direct Selection

In order to obtain specifically a cloned gene, it is essential to plate all the transformed cells on agar media, on which only desired recombinant clones can grow while the rest of the cells growth is totally inhibited.

To prove direct selection, a gene which specifies antibiotic resistance is cloned. As an example in the experiment a gene for kanamycin resistance from plasmid is cloned for direct selection. The *Eco*RI plasmid R6-5 in fact contains four different types of antibiotic genes (Tetracycline, chloramphenicol, kanamycin and sulphonamide). This plasmid can generate 13 fragments after digestion with endonucleases. To clone this gene *Eco*RI R6-5 plasmid cuts with endonucleases and generates 13 fragments. These fragments are ligated into the *Eco*RI site of pBR 322. Ligation gives 13 different recombinant DNA molecules. The transformed *E. coli* cells are plated on agar medium containing 50 μg/ml kanamycin.

Only recombinant molecules carrying kanamycin resistant gene can survive and are able to grow on the medium (Fig. 8.1).

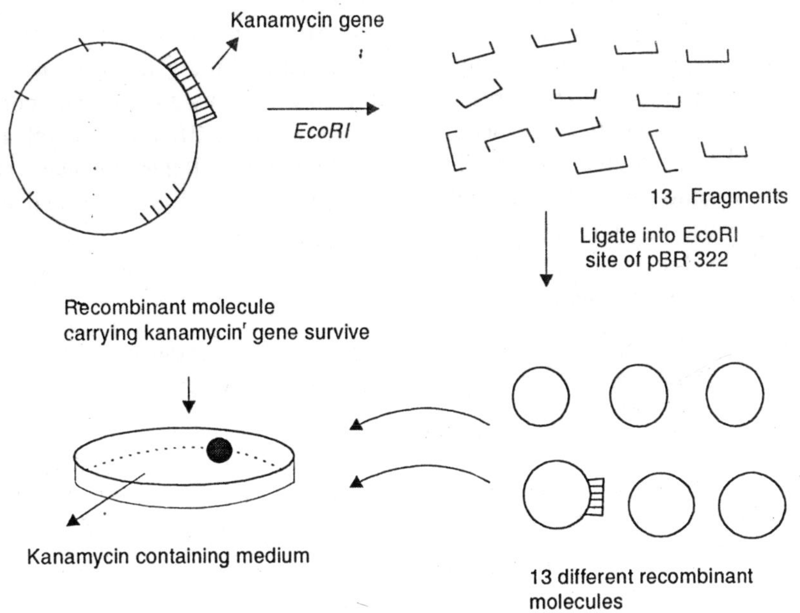

Fig. 8.1. Direct Selection

In another direct selection approach, identification of cloned gene is done by distinguishing wild and mutant bacteria by plating on tryptophan free media. The *E. coli* cell contains trp^A gene responsible for biosynthesis of tryptophan, while certain mutant strain contains non functional trp^{-A} gene and do not synthesize tryptophan. Hence, they can survive only if tryptophan is added to the medium and are called as auxotrophs. This mutant can be used to clone correct version of the trp^A gene. Total DNA is obtained by digesting with endonucleases, followed by ligation into vector, as a result many recombinant molecules are produced. In recombinant molecule trp^A gene is present. The mixture of recombinant molecules is used to transform into auxotroph *E. coli*: trp^{-A} cells.

After recombination majority of the transformants will be auxotrophs but few will have the correct trp A gene. All *E. coli* cell mixture are plated on media containing tryptophan. As predicted, only recombinant cells containing trp^A gene can survive and are able to grow in absence of tryptophan. Because trp^A gene can express (in absence of tryptophan) and produce tryptophan, an essential amino acid, whereas auxotroph cells cannot survive due to its non functional trp^{-A} gene in absence of tryptophan in the medium.

8.2 IDENTIFICATION OF CLONED GENE FROM A GENE LIBRARY

Gene library or genomic library is a collection of clones (genes) representing all the genes of an organisms. Gene library is prepared by isolation of total DNA from target organism subjected to partial digestion by restriction enzyme treatment and the resulting DNA is then ligated to vectors like bacterial plasmids, λ phage vector and yeast artificial chromosome (YAC). Comparatively it is feasible to construct genomic library for bacteria, fungi and yeast. The genomic library constructed for plants and animals contain innumerable number of clones, and it is extremely difficult to search for a cloned gene. A characteristic feature of all the multicellular organisms is the specialization of individual cells, for example, a plant contains leaf cells, petal cells, embryo cells and root cells, each containing the same complement of genes, but in different cell types, different sets of genes are switched on while others are silent.

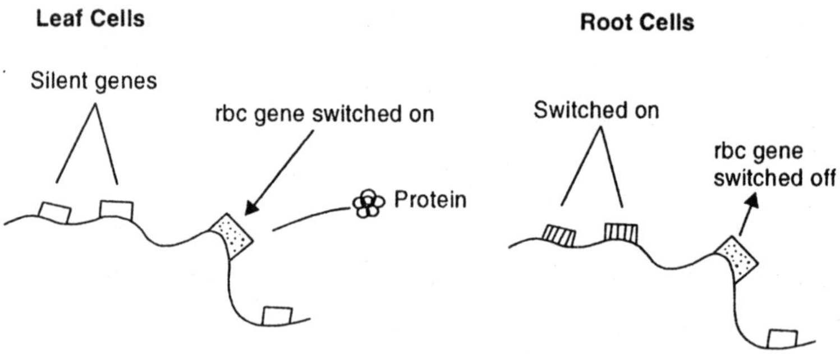

The gene rbc (ribulose bis phosphate carboxylase) is switched on in leaf cells and switched off in root cells.

8.3 CLONING STRATEGY—mRNA PREPARATION

The first step in cloning strategy is to obtain suitable source DNA. The source DNA for cloning can be chromosomal DNA or copy of a mRNA and complementary DNA (cDNA). The second step is primarily concerned with the production of DNA fragments. These fragments are prepared from chromosomal DNA digested with restriction endonucleases. The fragments can also be obtained from mRNA by using the enzyme reverse transcriptase to produce cDNA molecule. Each cDNA molecule is a copy of a single mRNA. DNA fragments are then inserted into appropriate vectors. This in turn can be introduced into population of bacteria. The vector with inserted fragments are introduced into the population of bacteria maintained on agar plate. Selection strategy can be applied at this juncture. Growth of transformed cells on media can be attributed to the fact that vectors carry antibiotic-resistance genes. Two genes namely tetracycline and ampicillin exhibit antibiotic resistance and can provide valuable direction for selection. Only bacteria containing plasmid will grow when plated onto agar containing the antibiotic. Each resistant bacterial cell grows to form a colony. Such a collection of cells containing foreign gene is known as gene library which represents every segments of chromosomal DNA or cDNA molecule from every different mRNA. The genomic library is a library without catalogue and can tell us which clone contains a particular target sequence. Screening of all the colonies for desired sequence is done in the final stage. Genomic library construction is described in the earlier chapter. To construct cDNA library total RNA is isolated and form this a fraction of mRNA is separated. Oligo dT is linked to cellulose material in chromatographic column which is used to purify mRNA from the rest of the total RNA. The enzyme reverse transcriptase converts mRNA to DNA (single stranded). At this stage it is RNA-DNA hybrid. The second strand of DNA synthesis is performed by DNA polymerase and completes strand synthesis. The RNase can recognise RNA-DNA hybrid molecule and digest RNA strand into small pieces. These RNA pieces can still able to hybridise with first cDNA strand and serve as a primer for DNA polymerase I, which uses original first DNA strand and synthesise second complementary strand.

8.4 IDENTIFICATION OF CLONE

8.4.1 Hybridisation probing

Nucleic acid hybridization can be used to identify a particular recombinant clone (gene) in a library using DNA or RNA probe complementary to the desired gene. First step in transfer cell colonies to nitrocellulose paper or nylon membrane. Alkanine treatment is done to denature DNA into single strand. This treatment ensures the retaining DNA on the nitrocellulose paper. This single stranded DNA is made to bind tightly to the nitrocellulose paper by a short period of heating at 80°C or irradiated with UV light if nylon membrane is used. The radiolabelled probe is denatured and applied to the membrane in a chemical solution that facilitates nucleic acid hybridization. The nitrocellulose or nylon membrane is washed to remove the excess of unbound probe. The position of the bound probe is then detected by autoradiography (Fig. 8.2).

8.4.2 Probing with non-radiolabelled method

Radioactive labelling methods usually create problems associated with disposal of radioactive waste which is hazardous to the researcher. An alternative version of non-radiolabelled biotic labelling method is useful. Biotin is an organic molecule binds with protein called avidin. After hybridization with the position of bound biotinylated probe it can be determined by washing with avidin coupled to a fluorescent marker (Fig. 8.3). In the second method of non-radioactive hybridization probing where the probe DNA is complexed with the enzyme horse radish peroxidase and is detected by degradation of luminol by enzyme with a emission of chemiluminouscence. The signal can be recorded on normal photographic film. Similar to autoradiography.

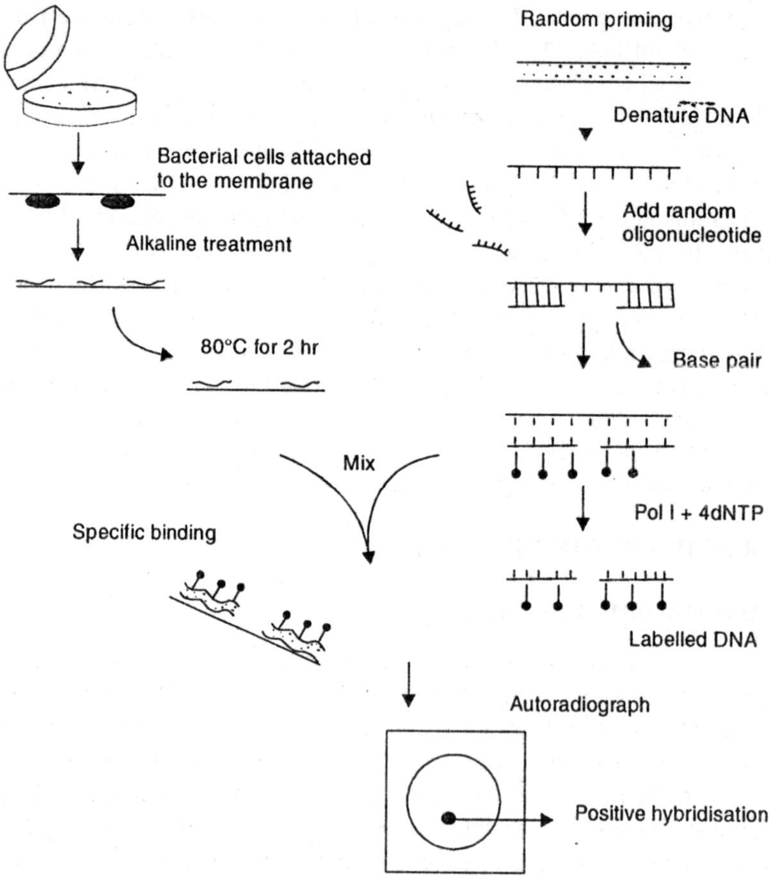

Fig. 8.2. Colony hybridization probing

8.4.3 Immunological Detection Method

Hybridization probing is done by immunological screening. In hybridization probing cloned DNA fragment is directly identified where as in immunological method identification is based on translated product of cloned gene.

The translated protein is identified, purified and is injected into the rabbit. The antibody produced in the rabbit blood stream is isolated and labelled. Meanwhile recombinant colony are transferred to polyvinyl membrane. The cells are lysed and specific labelled antibody is added. Since labelled can be radioactive colonies that bind to the label are detected by autoradiography.

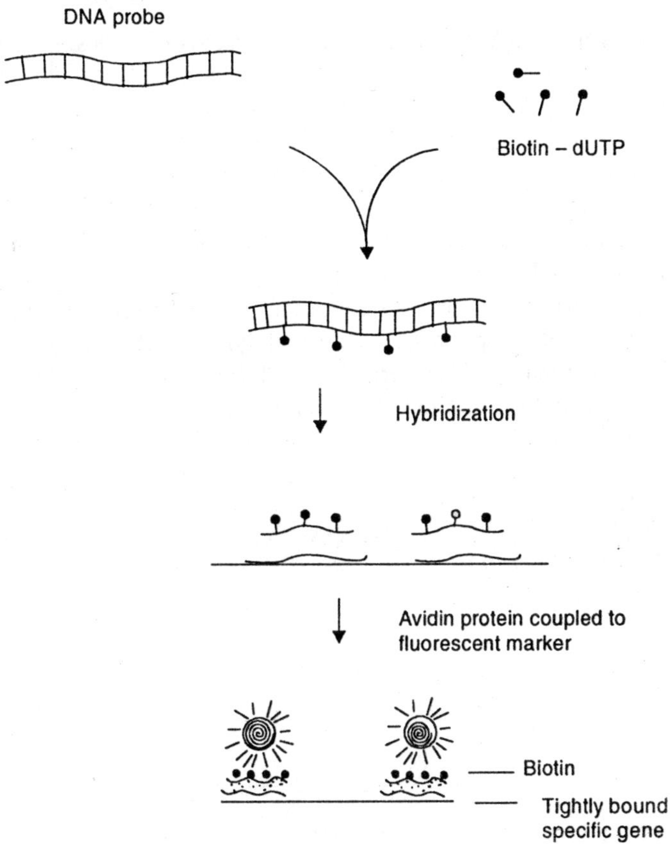

Fig. 8.3. Non-radiolabelled probe

8.5 STUDYING GENE LOCATION OF SMALLER SIZE

Several techniques are available for determining the location of a cloned gene on a DNA molecule. Locating the position of a cloned gene on a small DNA molecule is done by Southern hybridization technique. For example, to identify kanamycin antibiotic "r" gene. DNA of *E. coli* R6-5 is removed and *Eco*RI restriction digestion is done. The resulted fragments are then electrophoresed. One of the separated DNA fragments on a gel contains antibiotic resistance gene. This fragment will be the same as that in recombinant pBR 322 molecule that carries the same antibiotic 'r' gene.

The DNA fragments are transferred to nitrocellulose membrane by southern hybridise technique. When DNA fragments binds to nitrocellulose membrane they can be made single strand by alkali treatment. The labelled probe is applied to facilitate hybridization and is followed by autoradiography. This will show which DNA fragment contain the cloned gene (kanamycin resistant gene).

8.6 STUDYING GENE LOCATION OF LARGER SIZE

Higher eukaryotic DNA contains several thousands of genes, organised on several chromosomes. Identification of a particular gene in a huge size of the chromosome is a mammoth task. Southern hybridization is feasible for DNA of plasmid, bacteriophage and virus. This technique cannot be extended to locate cloned gene on large DNA molecule specially when molecules is more than 250 kb in size. Therefore several other methods have to be used to locate position of cloned eukaryotic genes on chromosomal DNA molecule.

To find out which chromosome carries the target sequence. Special type of southern hybridization is carried out to separate chromosomal DNA molecule by gel electrophoresis. A special type of gel electrophoresis is known as orthogonal Field Alternative Gel Electrophoresis (OFAGE). It involves electric field is applied alternatively between the two points of the electrode and each pair is set at an angle of 215° to the length of the gel.

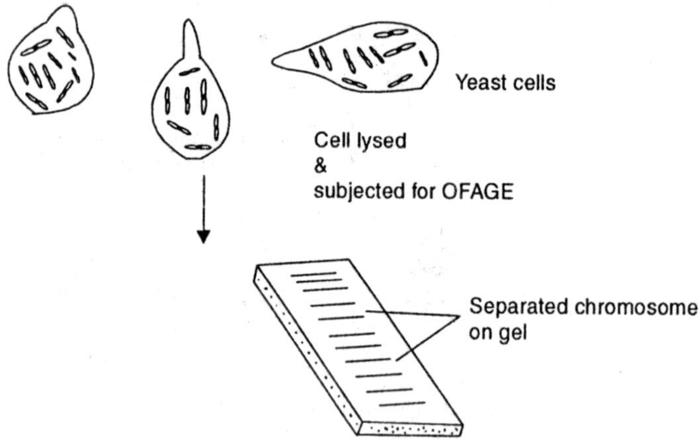

Fig. 8.4. Separation of chromosome by OFAGE technique

Electrophoretic separation of chromosome on the gel is followed by shifting of chromosome from gel matrix to nitrocellulose paper by southern transfer and studied by nucleic acid hybridization probe. A cloned gene can be identified (Fig. 8.4) by using this technique chromosome that carries.

8.6.1 Insitu Hybridization

The mammalian and other higher eukaryotic system contains larger DNA molecule. Some times larger than 50,000 kb. Hence, non-conventional methods like gel electrophoresis is not feasible. The gene location on these larger molecule can be identified by insitu hybridization which can prove accurate information regarding the position of cloned gene. Insitu hybridization involved standard light microscopy used to observe chromosome in a cells that are in the process of division. This technique enable us to directly visualise location of a cloned gene in a light microscopic image of the chromosome. In the process of in-situ hybridization cells are first treated with fixatives and attached to a glass slide and then incubated with sodium hydroxide and ribonuclease to denature DNA and degrade RNA. Due to alkaline treatment base pairing between the individual polynucleotide strand is broken. A complementary strand of cloned gene is labelled and mixed with chromosome preparation on the slide. When probes make a contact with denatured DNA within the chromosomes, hybridization takes place between cloned gene and chromosomal copy. A dark spot on the X-ray film appeared when subjected for autoradiography. The position of dark spot indicated the location of a cloned gene in the chromosome. Insitu

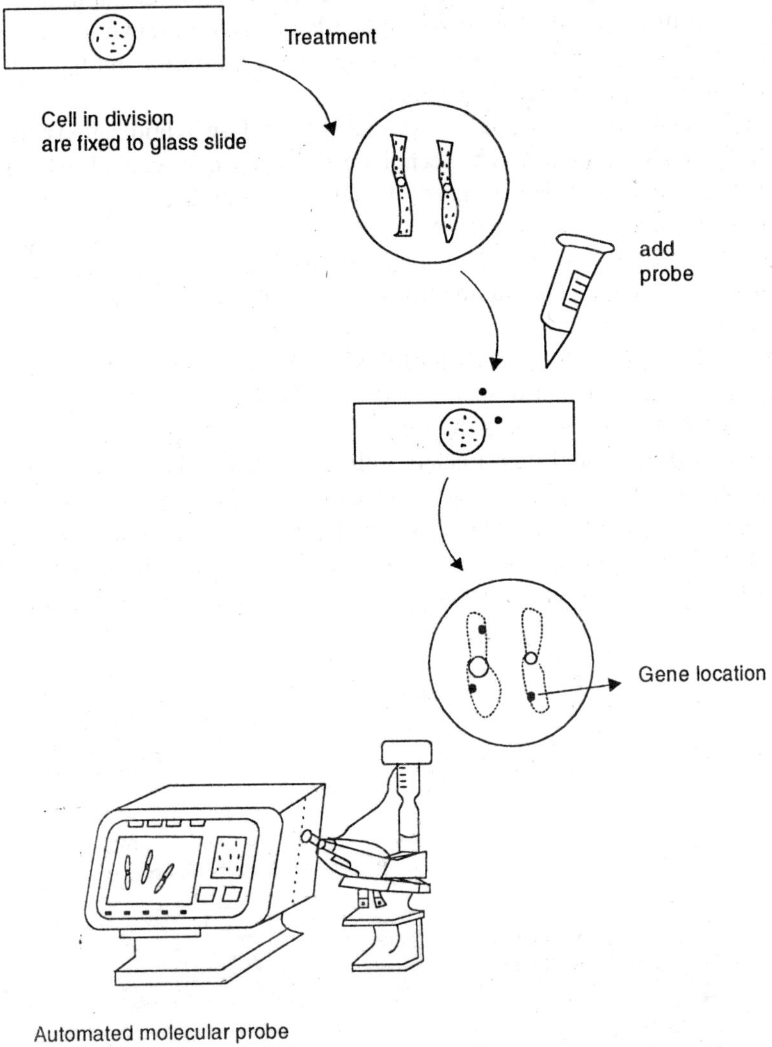

Fig. 8.5. In situ hybridization

hybridization with radioisotope labelling is comparatively different. It is very useful in locating the position of number of genes. An alternative radio active probe is fluorescent marker. It can be attached to the probe and mixed with chromosomal DNA to facilitate hybridization process. It can be observed directly by specially designed microscope (Fig. 8.5).

8.7 STUDYING GENE EXPRESSION

The first process in gene expression is the transcription of gene into a complementary RNA and translated to produce protein. Some of the genes when expressed produce tRNA or ribosomal RNA. In order to study the gene expression, RNA transcript must be studied. The fate of introns, exact locations of the start and end point of the transcription is important. The expression of a cloned gene to be studied is otherwise known as transcript analysis are based on hybridization between RNA transcript and fragment of DNA containing corresponding gene.

Electron microscopy can be used primarily to analyse hybridization between different DNA molecules. The technique has been now extended to study DNA-RNA hybrids. It is very useful to determine introns, if a gene contains it. It has no homology with RNA transcript. Due to non base pairing they loop out as a characteristic appearance. The loop out structure can be seen by electron microscopy. The number of loop can indicate number of introns in a gene.

In another approach on studying DNA-RNA hybrid. It involves a single-stranded specific nuclease such as SI. This enzyme can degrade single-stranded DNA and RNA including intron region. If a DNA containing gene is hybridised to its and then treated with S1 nuclease. The non hybridized single strand DNA region at each end will be digested, including introns. The result is completely double stranded RNA-DNA hybrid. The single stranded DNA fragments can be protected from S1 nucleus. If RNA strand is degraded by alkaline treatment the sizes of the protected DNA fragments

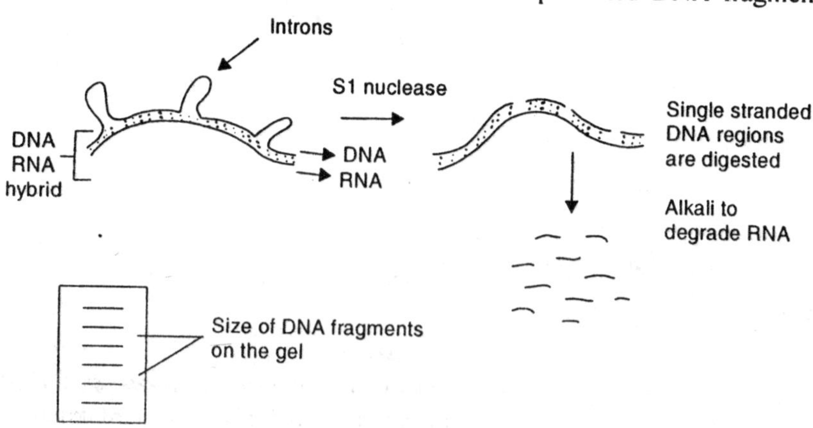

Fig. 8.6. Studying gene expression

could be measured by gel electrophoresis. This technique can determine precise start and end point of the transcript (Fig. 8.6).

8.8 STUDYING THE REGULATION OF GENE EXPRESSION

Differential gene expression is a unique phenomenon during growth and development of an organisms. Number of factors are responsible for regulation of gene expression. A simple gene expression can be seen in bacterium *E. coli*. Some genes are expressed, if cell need to produce some substance. The genes of certain kind is switched off or never expressed when some substance is present or just appears the classic example is regulation of trp operon in *E. coli*.

When tryptophan amino acid is present or just added to the medium trp gene will automatically switched off. If tryptophan is withdrawn from the medium within short period of time trp operon may switched on to bio-synthesise tryptophan. Studies have shown that there are many control sequences in the DNA to which certain proteins can bind and regulate gene expression.

It has been shown that a regulatory protein may repress gene expression in which gene is switched off—when protein binds to the control sequences. Hence, gene regulation studies involves locating control search and its role in gene expression.

8.8.1 Identifying protein-binding site

A control sequence is a region of DNA that can bind a regulatory protein. There are two approaches to identify the control sequence. In the first approach a protein bound to DNA fragment is identified by virtue of its increased molecular of mass. The DNA fragment containing a bound protein is identified by gel electrophoresis. The DNA complexed proteins moves slowly during electrophoresis than uncomplex of DNA molecule. This is known as gel retardation. Further, the control sequences is digested with restriction endonucleases before mixing with the regulatory proteins. The location of the control sequence is then determined by finding the position on restriction map of the fragment that is retarded during electrophoresis.

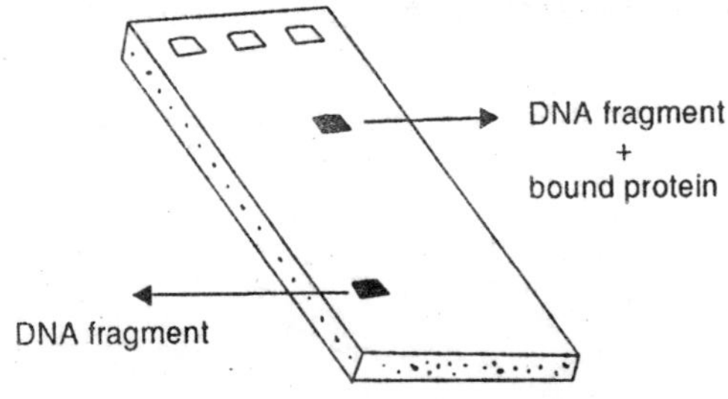

Fig. 8.7. Identifying protein binding site

8.9 STUDYING THE TRANSLATED PRODUCT OF A CLONED GENE

In recombinant DNA technology identification of translated product of a cloned gene in feasible at specific points. Two important techniques like hybrid-release translation (HRT) and hybrid arrest translation (HART) are widely used to identify translated product of a cloned gene. These techniques depends on the availability of purified specific mRNA and direct the synthesis of proteins in cell-free translation system. The mRNA sample is mixed with 20 different amino acids and one of the amino acid is radiolabelled, the mixture is added to cell-free translation.

The mRNA in cell free system in translated to protein. These are then separated and removed after electrophoresis. HRT and HART system were found to be suitable when cDNA clone is prepared directly from mRNA sample. Using HRT system to identify particular translational product, cDNA is denatured, immobilised on a nitrocellulose membrane and incubate with available specific mRNA. The cDNA is hybridised with its complementary mRNA. After discarding unbound molecules, the hybridised mRNA is removed and translated in cell free system. In HART system hybridization takes place between mRNA and cDNA. After hybridization the unbound mRNA is not discarded instead the entire sample is translated in the cell free system. The hybridization of mRNA in these case is not get translated due to its hybridization nature. So all the proteins except the one coded by the cloned gene is translated. The cloned genes translated product is therefore identified as the protein that is absent from the autoradiograph.

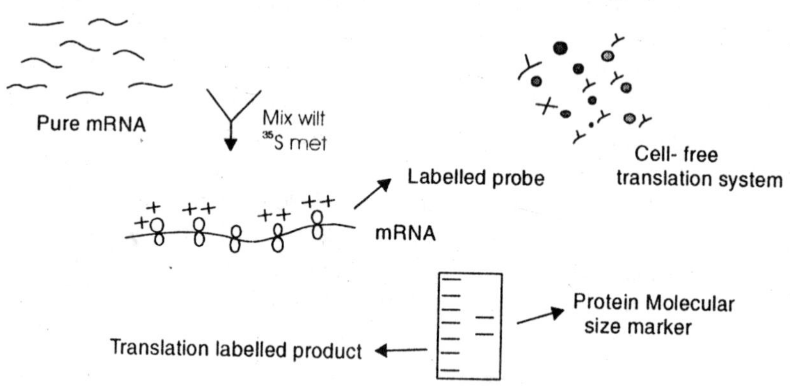

Fig. 8.8. Identification of translated product of a gene.

9

Gene Cloning in Research of Biotechnology

9.1 PRODUCTION OF PROTEINS FROM CLONED GENES

One of the most remarkable achievement in recombinant DNA technology is the successful cloning the expression of genes. Rapid growth of biotechnology industry is because of development in gene cloning technology. The economic growth of several major pharmaceutical companies in western countries is due to effective implementation of gene cloning technology. Many useful recombinant products have already been produced using gene cloning technique. Recombinant DNA technology has made a quantum shift in the production of recombinant pharmaceutical proteins in tiny microorganisms. Although breath taking progress have been made in the field. Several lacunae such as designing of special cloning vectors, satisfactory yield of proteins has to be fulfilled.

9.2 DESIGNING OF EXPRESSION VECTOR

In gene cloning strategy if foreign gene is ligated into the vector and cloned in *E. coli*. It may not produce sufficient amount of foreign proteins. The signal system surrounded the gene can play a decisive role in optimising gene expression. There are three important signals for *E. coli* genes.

(a) The "promoter" which initiate transcription by facilitating binding of sigma subunit of RNA polymerase.

(b) The "terminator", which marks the point at the end of the gene where transcription terminates. A terminator is usually a nucleotide sequence that can base pair with itself to form a stem-loop structure.

(c) The ribosome binding site, it is a short nucleotide sequence recognised by ribosome as the point at which it can attach to the mRNA molecule.

The genes of higher organisms are surrounded by several expression signals. Their nucleotide sequences are distinguished remarkably from prokaryotic sequence. If a foreign gene is inserted in *E. coli*. It is unlikely that bacterial RNA polymerase would recognise human promoter and initiate transcription. To overcome this problem foreign gene can be cloned in bacteria and placed under the control of bacterial promoter and termination signal. The cloning vehicle which provide three signals can be efficiently used in the production of recombinant protein are called expression vector (Fig. 9.1).

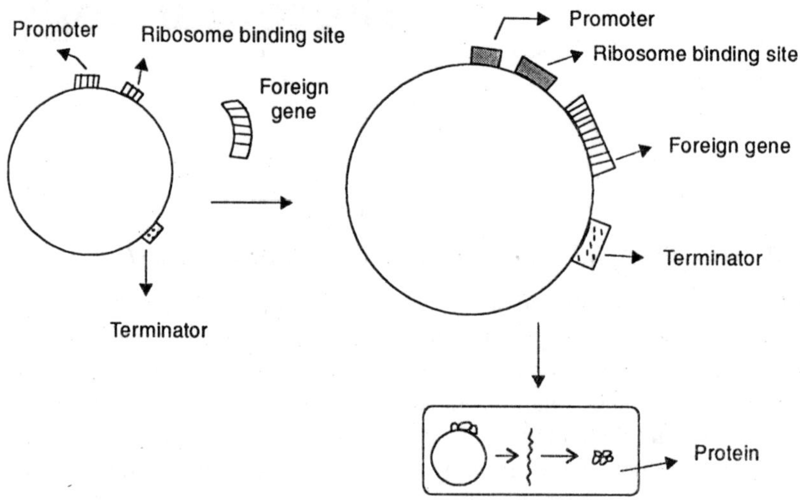

Fig. 9.1. Construction of expression vector

In order to optimise gene expression for the production of sufficient amount of recombinant proteins. It is essential to place strong promoters before a foreign gene. Strong promoters are those that can sustain high rate of transcription. Strong promoters are those usually control genes whose proteins are required in large amounts by the cell. Whereas the weak promoters are relatively insufficient, direct transcription of genes whose products are required in small amount by the cell. The expression vector should carry strong promoter to transcript at higher rate (Fig. 9.2).

Another important factor that govern optimization of gene regulation is induction and repression. Any inducible gene is one whose gene is switched on by the addition of chemical to the growth media. In contrast a repressible gene is switched off by the addition of regulatory chemical.

9.3 PROMOTER FOR *E. coli*

Most frequently used promoters in expression vectors are (a) the lac promoter: which control transcription of lac Z gene coding for β-glactosidase. This gene is carried by vector like pUC and M13. The lac promoter is induced by IPTG (Indodyl galactoside)

(b) The trp promoter: Involves control of cluster of genes coding for several enzyme in biosynthesis of tryptophan (c) λ P_2 promoter: This is responsible for transcription of the λ DNA. This is a strong promoter recognised by bacterial RNA polymerase.

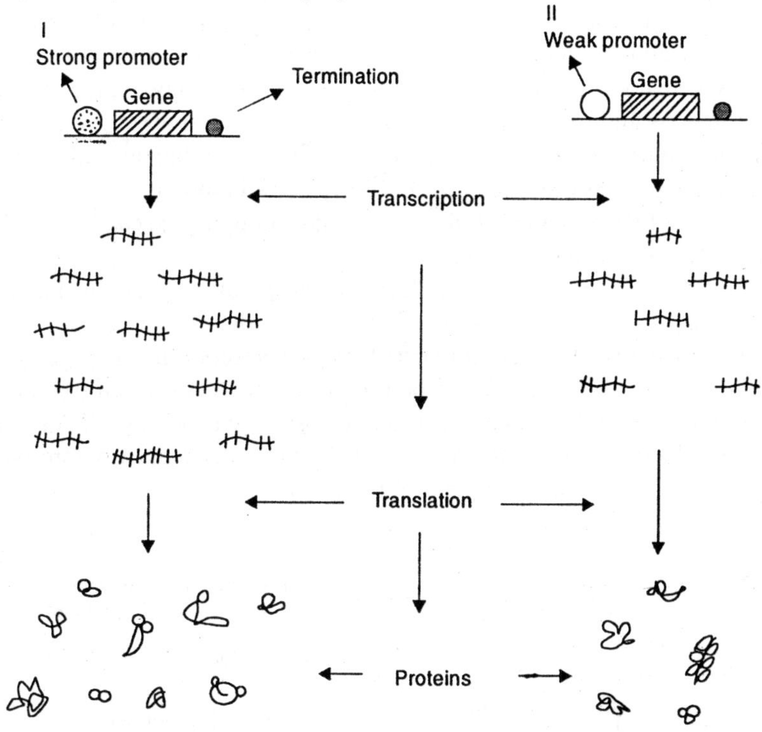

Fig. 9.2. Efficacy of strong promoter in transcription

9.4 PROTEIN PRODUCTION BY REGULATABLE PROMOTER

One of the most widely used strong promoter for protein production in microbial system is trp (tryptophan) gene (operon) and P_L promoter. Increasing production of proteins can be achieved by the use of temperature sensitive P_L promoter and trp gene (operon) regulatable promoter. The P_L promoter and CI repressor protein are used in *E. coli* system. It is used for the controlled production of foreign protein for limited scale. The P_L promoter is controlled by the CI repressor protein of λ bacteriophage. When temperature is below 30°C (<30°) CI protein is active and does not allow the p_L promoter to direct transcription. In a small fermenter, where cells carry plasmid strain which was first grown at 28°C. At lower temperature CI repressor is active and P_L promoter is turned off. At this stage maximum cell growth is achieved. Later, when temperature is shifted to 42°C CI repressor in inactive and do not interfere with P_L promoter at high temperature. Hence, when temperature of the culture is shifted to higher side (>42°) foreign gene placed under the control of P_L promoter is expressed and produce proteins. Thus by altering temperature in a fermentor proteins can be produced by regulating promoter.

In pilot scale culture, it is possible to produce increased amount of proteins by using trp promoter. Earlier method of temperature manipulation is not feasible as economically costlier. Therefore in large scale culture more than 250 litres of culture can be maintained in bioreactor. Two different types of plasmids were constructed in the laboratory. One is low copy number plasmid and another is high copy number plasmid. The CI repressor was placed under the control of trp promoter in a low copy number plasmid. In the second type of plasmid, foreign gene is placed before P_L promoter. By utilising these two laboratory designed plasmids foreign gene can be switched on or off by monitoring tryptophan level.

If tryptophan is withdrawn or reduce its concentration in the medium, trp promoter is turned on and transcribe CI gene to produce CI repressor protein. This protein binds to P_L promoter and block transcription of target gene. Therefore foreign protein will not be produced. However, when tryptophan level is raised in the medium trp promoter is turned off and CI repressor protein production is blocked. As a result P_L promoter placed before target gene is turned on and allow target or foreign gene to transcribe and translated to produce proteins (Fig. 9.3).

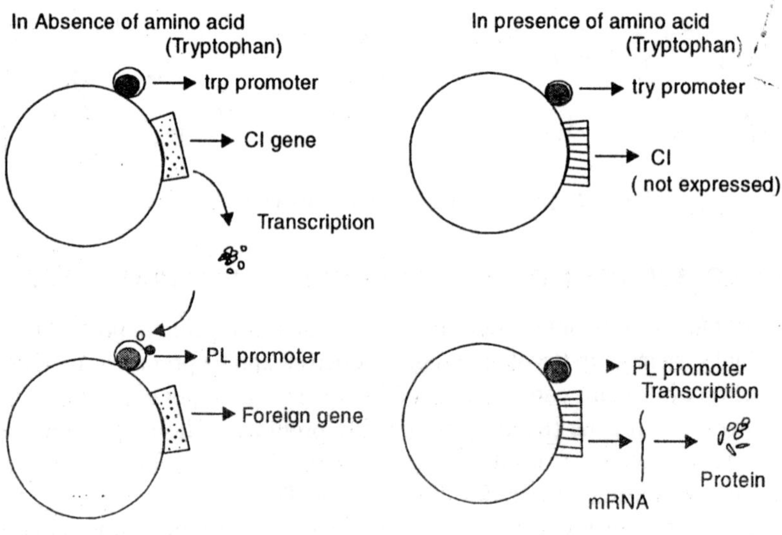

Fig. 9.3. Control of gene expression by tryptophan amino acid.

9.5 FUSION PROTEINS

The *E. coli* and in other bacterial system it has been observed that only minute quantities of foreign protein get accumulated. This low level accumulation is probably due to degradation by bacterial proteolytic cleavage enzyme. When a foreign gene of eukaryotic origin is cloned in bacterial (prokaryotic) cell, the expressed product recognised as alien protein result in proteolytic cleavage and destroy it. In order to save foreign protein from bacterial enzyme attack it is essential to modify strategy by attaching foreign gene to one of the bacterial gene. This novel gene construct is then ligated into vector and cloned in bacteria. When a fusion vector system is directed for transcription, expressed gene is translated and produced as a fusion protein. Although fusion protein was found to be resistant to cleavage enzymes. Several clinical trials was found to be unsuitable and may alter biological functions of protein. Therefore it is essential to remove unsuitable protein sequence (bacterial) from the target protein. This can be made at DNA construct level to facilitate cleavage of unwanted protein from target protein. In a novel approach protease recognition sequence is attached between target and bacterial gene. When bacterial cell produced fused protein. They are treated with specific protease to facilitate proteolytic cleavage of amino acid sequence present between bacterial and target protein.

9.6 PRODUCTION OF RECOMBINANT PROTEINS BY EUKARYOTIC CELL

Efficacy of lower eukaryotic organisms like yeast and other filamentous fungi for the production of several recombinant proteins have been realised. They can be efficiently used for protein production than its counter part *E. coli*. Several advantages are behind the logic of selection of yeast as a cloning vector. The protein production in yeast and fungi are closely related to animals. They can perform certain post translational modifications which are absent in prokaryotic bacterial system as an added advantage these can be cultured in fermentors like bacteria and enable us to produce recombinant protein.

The yeast *saccharomyces cerevisiae* is commonly used as a popular vector for protein production. The alien gene is often placed under the control of GAL promoter. The GAL promoter can undergo regulation by galactoside. Despite higher production of recombinant protein in yeast, it fails to glycosylate animal protein correctly. It also lack proper secretary mechanism. Inspite of certain disadvantages yeast is still to be an ideal choice because it is a safest organism for protein production and also scientist have clear knowledge of its genetic material. In recombinant DNA work it is possible to clone a large segment of DNA in *saccharomyces cerevisiae*, which is then maintained as a separate chromosome in the host yeast cells.

The two most popular filamentous fungi are *Aspergillus nidulans* and *Trichoderma reesei*. These are the organisms having advantage of glycosylation of animal proteins and ability to secrete protein. Insect cells provide a important alternative to mammalian cells for animal protein production. They can provide substantial amount of recombinant protein. The expression system is based on baculovirus. These are common in insect but do not infect vertebrates. The virus contains polyhedrin gene. This gene products accumulates in insect cell as large number of inclusion bodies. It can occupy upto 80% of the total cell protein. Similar level of protein production takes place if the normal gene is replaced by foreign gene.

QUESTIONS

1. What are expression vectors? Describe methods of production of foreign proteins.
2. Describe how a clones gene is regulated by trp promoter.

SHORT QUESTIONS

1. Fusion proteins (BU April 2002)
2. Regulatable promoters
3. trp operon

10

Recombinant DNA Technology in Medicine

The remarkable achievements in all areas of molecular biology, artificial gene synthesis, enzymology of recombinant process and an understanding of gene expression, have revolutionised research in medicine and pharmaceutical industry. Cloning of transgene and production of proteins of importance is a remarkable success in medicine. Today, proteins are produced through recombinant DNA technology for treatment of cancer, allergies, heart attacks and infections. Several attempts have already been made through recombinant technology for the development of new generation of drugs for therapeutic use.

In order to produce recombinant drugs several expression systems are developed. The most popular expression systems are the bacteria *E. coli* and *Bacillus subtilis*, yeast and mammalian cells. Microorganisms like bacteria offer simplicity and large yield production with low costs. Following are some of the important recombinant products produced through this cutting edge technology.

10.1 PRODUCTION OF RECOMBINANT INSULIN

The first recombinant licensed drug produced through genetic engineering was human insulin. Insulin regulates sugar metabolism. Any inborn error may result in diabetes. If the pancreas fail to synthesise insulin due to the damage of the cells of pancreas or any other reason, then external supply of insulin is required to regulate the sugar metabolism. Hence, there is a huge demand for this hormone throughout the world. Insulin is obtained from pancreas of slaughtered animals. However, once it is administered some people develop serious immunological reaction against animal insulin. Human insulin produced through recombinant DNA methods are safe and reliable. Insulin is made up of two peptides interlinked by disulphide linkage.

The synthetic human insulin genes were synthesised for A and B chain peptides and are separately inserted into the plasmid of *E. coli* (expression vector). The chimeric gene construct, which includes insulin and β-galactosidase sequences are able to express and produces as fusion protein. The fusion protein is protected from the bacterial proteases β-galactosidase chain (prokaryotic origin). The chain A and chain B proteins were purified from bacterial extract and insulin chains were released by treatment with cyanogen bromide to cleave peptide bond after methionine residue, the two chains are then purified, mixed to form active recombinant insulin (Fig. 10.1).

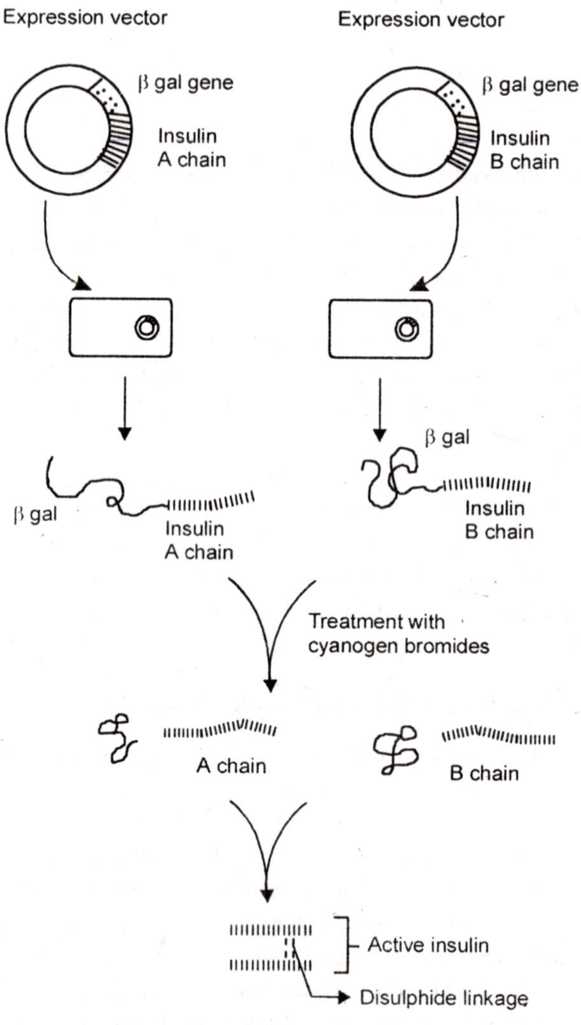

Fig. 10.1. Production of human insulin in *E. coli.*

10.2 RECOMBINANT HUMAN GROWTH HORMONE (HGH)

Human growth hormone (hGH) is synthesised in pituitary gland and regulate growth and development. Any deficiency in hormone production results in dwarfness. Treatment with animal hormone on the other hand are ineffective in reversing dwarfism. Some good effort have been made to extract hGH from human cadaver and administered to children. They developed with serious viral infection. Hence, hGH produced by recombinant technology provided a safe drug to reverse dwarfism.

Production of hGH could be achieved by fusing hGH coding sequence to bacterial signal sequence. The DNA fragment encoding a bacterial signal sequence is placed before hGH sequence. The presence of bacterial signal sequence ensures secretion of protein which gets accumulated in periplasmic region. These secreted proteins can then be released for extraction by hypotonic disruption of outer membrane.

In the periplasmic region, signal peptide is removed by bacterial proteases. Thousands of children are benefitted from administered hGH (Fig. 10.2).

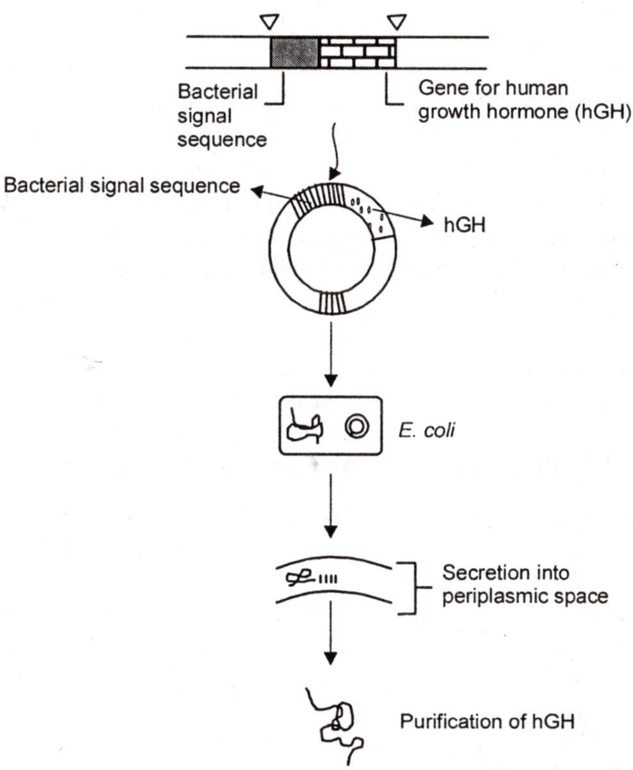

Fig. 10.2. Production of recombinant growth hormone (hGH).

10.3 PRODUCTION OF RECOMBINANT VACCINES

One of the highlights of modern medicine in the field of new generation drugs is the production of recombinant vaccines. Vaccines produced through conventional methods are of two types *individual vaccines* and *attenuated vaccines*. Inactivated vaccines are chemically or heat killed infectious organism. Attenuated vaccines are live infectious viruses or bacteria but, they are altered in such a way that they are not able to multiply in inoculated organism. Both the types of above said vaccines could provide surface antigenic proteins to immune cells (lymphocytes). They respond immediately by producing antibodies and destroy infectious organisms. However, these types of vaccines may continue to have contamination of organisms which are potentially dangerous. To overcome this problem recombinant vaccines can be used which are reliable and safe. Since recombinant vaccines are only antigenic proteins, devoid of genetic material is safe for therapeutic use.

Recombinant vaccines have been commercially produced for hepatitis B virus in yeast and in animal cells.

The general method for recombinant vaccine production in yeast cell is as follows:

Hepatitis B virus contains surface protein envelopes enclosing small 3.2 kb DNA that has been sequenced completely. To prepare a vaccine against hepatitis B virus, hepatitis B virus surface antigen (HBsAg) gene was cloned into a high copy number vector for expression in yeast. The expression vector which contains a strong yeast promoter of alcohol dehydrogenase was placed before HBsAg gene. A transcription terminator was placed down stream. The vector contains replication of origin and marker for both yeast and bacteria. Yeast transformed with the plasmid produced nearly 50-75 mg of the viral proteins in larger fermentor. The purification HBsAg protein was found to aggregate into particles (Fig. 10.3).

The recombinant vaccines by trade name 'Shanvac' have been commercially produced by Hyderabad based company Shantha Biotech.

10.3.1 Sub Unit Vaccines for Foot and Mouth Disease

Foot and mouth disease virus (FMDV) cattle is controlled to some extent by conventional vaccines that provide viral protein to immune cells for the production of antibodies.

In genetic engineering capsid viral protein cDNA was cloned in *E. coli* expression and vector expressed as fusion protein. The fusion protein containing viral protein fragment was able to synthesise neutralising antibodies against FMDV.

10.3.2 Peptide Vaccines

Peptide vaccines are small portion of protein (domain) which can effectively function as a sub unit vaccine and are able to generate neutralising antibody.

These short peptides, that mimic may have been responsible for immunogenic responses and could be used as peptide vaccines. The classic example is the use of chemically synthesised domains of FMDV viral proteins. They were tested as potential peptide vaccine. The domain regions are located near the C and N terminal ends of viral protein. After a series of tests, a single inoculation with peptides of amino acids of 141 to 160 are sufficient to generate neutralising antibody against foot and mouth disease virus. In USA the clinical trials are undertaken on peptide vaccine production based on epitope of plasmodium *Falciparum circumsporozoite* protein.

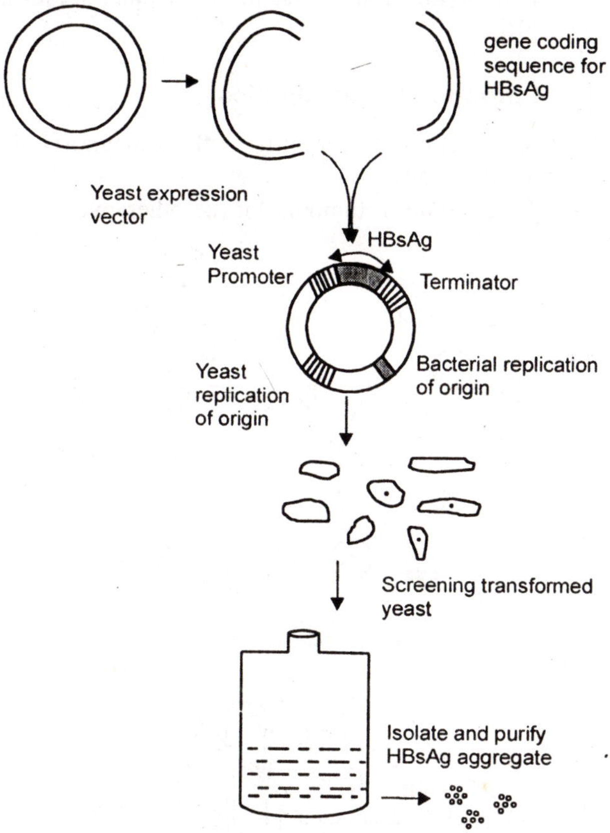

Fig. 10.3. Production of recombinant vaccines.

10.4 MONOCLONAL ANTIBODIES

The onset of hybridoma technology have made significant contribution in the production of specific antibodies. These are commonly known as *monoclonal antibodies*. In modern medicine, monoclonal antibodies function as magic bullets. These magic bullets can effectively be used to control infectious agents and tumour cells. One of the main strategy of hybridoma technology is that, a cell should produce substantial amount of monoclonal antibodies by retaining capacity of immortality. To obtain hybridoma cell, B lymphocytes and cancerous myolema cells were selected for fusion purpose. These two cell types exhibit unique character of antibody production (B cells) and continuous cell division (myolema cells).

The mouse is inoculated with an antigen (Ag) of interest. This stimulates the proliferation of lymphocytes expressing antibodies against injected antigen. Lymphocytes are taken from the spleen and fused with myeloma cells by treatment with polyethylene glycol (PEG). Hybrid cells are selected by growth in HAT medium and transferred to culture in microtitre wells for several days (Fig. 10.4).

Applications of monoclonal antibodies

(i) Monoclonal antibodies are currently used for diagnosis of infections.
(ii) Used in tumour therapy.
(iii) Also used for imaging of tumours for radiotherapy.

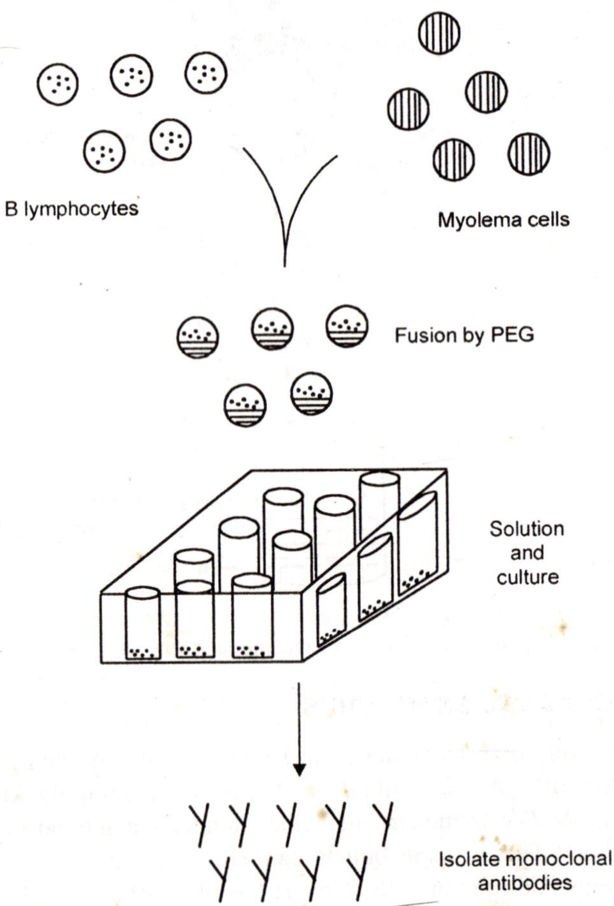

Fig. 10.4. Production of monoclonal antibody.

10.5 DIAGNOSTIC REAGENTS

Traditionally, diagnostic reagents have three forms

(a) Biochemical reagents for assaying specific enzyme.

(b) Immunological diagnostic system.

(i) Antibodies for detecting specific proteins by Radio Immuno Assay (RIA) and Enzyme Linked Immunosorbent Assay (ELISA).

(c) A new class of diagnostic reagents which are DNA probes or RNA probes. (DNA Diagnostic system).

We are describing only immunological and DNA diagnostic system to fulfil the scope of this book.

10.5.1 Immunological Diagnostic System

The technique is very useful for clinical microbiologist to detect pathogenic organisms. For this reason, many monoclonal antibodies and polyclonal antibodies have been developed for use as immunodiagnostic agents through recombinant DNA technology.

Production of Antibodies in *E. coli*

Antibodies can be produced in *E. coli* through recombinant technology as follows:

(i) Isolate mRNA from human antibody producing B lymphocytes.

(ii) Convert mRNA to cDNA using reverse transcriptase.

(iii) Amplify heavy and light chains of antibody gene by PCR. The amplified cDNA preparation is then treated with specific restriction endonucleases and cloned into bacteriophage λ vector.

(iv) More specifically heavy chain sequence is cloned into bacteriophage vector namely heavy chain, and also light chain sequence is cloned separately into bacteriophage A vector namely light chain B.

(v) When heavy and light chains are expressed during the lytic cycle of the bacteriophage λ. Combine heavy and light chains into one bacteriophage vector.

(vi) Screening for plaques is done for antigen binding.

(vii) Harvest antibodies from the cultured organisms. The techniques of monoclonal antibody production have already been discussed in this chapter.

10.5.2 DNA Diagnostic Reagent

The sequence of nucleotides, if labelled can be effectively used as diagnostic determinants. Generally, these diagnostic determinants are called DNA probes or RNA probes, more precisely nucleic acid probes. For labelling the probe, commonly phosphorus-32 is used.

Preparation of DNA probe:

The DNA probe is prepared by random primer method as follows:

(i) In double stranded DNA containing the sequence that is to act as the probe is denatured and an oligonucleotide sample containing all possible sequences of six nucleotides is added. (It is statistical certainty that some of the molecules of the oligonucleotide mixture will hybridize to the unlabelled, denatured probe DNA).

(ii) In the presence of klenow fragment and four deoxyribonucleotides, one of the four deoxyribonucleotides is labelled.

(iii) The bound oligonucleotides act as primers for DNA synthesis.

(iv) The synthesised DNA is labelled and can be used as a probe to detect the presence of a complementary DNA sequence in a source DNA sample (Fig. 10.5).

There are at least two possible sources of probes: one is from cloned DNA, second is the nucleotide sequence of a synthetic probe is based on the probable nucleotide sequence that is deduced from the known aminoacid sequence of the protein encoded by the target gene.

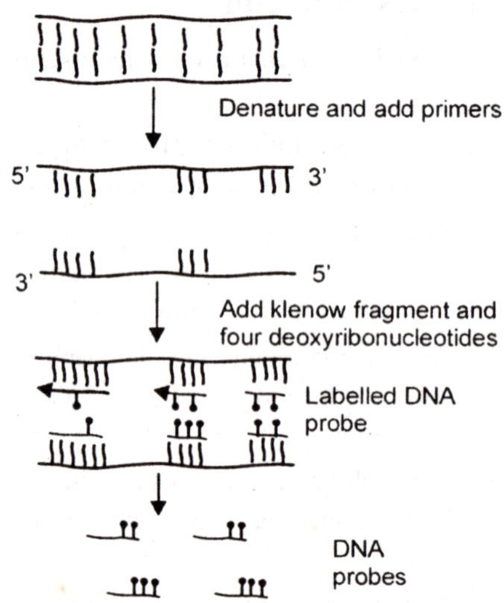

Fig. 10.5. Preparation of labelled probes.

Nonradioactive Hybridization probe

Nucleic acid hybridization probe as a diagnostic reagent, contains radiolabelled phosphorus 32. It is short lived and dangerous. Hence, safety guidelines are required. Alternately a nonradioactive system for signalling hybrid DNA have also been developed. It can be prepared in the laboratory as follows:

(i) The biotin-labelled probe is hybridised to the target DNA.

(ii) The chicken egg white protein or streptavidin is added.

(iii) A biotin-labelled enzyme, alkaline phosphatase or peroxidase is added.

(iv) Chromogenic or chemiluminescent substrate is added, the former causes a change in the colour and the latter produces change or the light produced as it converts substrate to product that is measured by absorbency.

Biotin-labelled DNA is stable for one year and used as potential molecular diagnostics.

10.6 GENE THERAPY

Gene therapy is a genetic engineering technique of correcting a defective gene in human beings. During the growth and development of an organism, all the genes are committed for differential expression. Specific proteins have to be synthesised at a specific time during growth and development. A minor defect in the gene results in the production of architectural abnormalities of a protein. Any altered protein may result in altered phenotypic character. An altered sequence in a gene is known as "defective gene" or "mutated gene". Any altered protein structure or activity of an enzyme, might cause either an accumulation of a substrate that are toxic to a cell or any deficiency of a compound that is vital for normal activity of a cell.

As the advancement in recombinant DNA technology is progressing rapidly, an attempt has been made in 1980 to replace a defective gene, which is popularly called as "Gene therapy" or "Gene replacement therapy". However, first clinical trials of gene therapy for Adenosine deaminase deficiency began in 1990. Gene therapy in future may have a potential answer to various single-gene human disorders. Currently gene therapy for ten single gene human disorder is in clinical trials.

10.6.1 Types of Gene Therapy

Generally gene therapy involves two types; one is somatic cell gene therapy and the other is germ line gene therapy, which involves selection of reproductive cells. Most of the work is currently being carried out offering somatic cell gene therapy. However, many gene defects have also been corrected using germ line therapy. In germ line therapy the fertilised egg is used to inject plasmid containing remedial gene (corrected or replaceable gene) or remedial gene is inserted in retroviral vector and transferred into fertilised egg. The amount of mutagenesis caused by retroviral insertion may be unacceptably very high. According to many individual germ line theory it is morally, ethically not accepted as it involves the transfer of manipulated gene to the children from treated patient.

10.6.2 Methods of Gene Therapy

Somatic cell gene therapy is classified into the following methods — Exvivo gene therapy, Invivo gene therapy and Antisense therapy.

10.6.2.1 Exvivo gene therapy

The first step in the exvivo gene therapy includes:

(a) Isolate cells with a defective gene from an affected individual.
(b) Growing the isolated cells in culture.
(c) Correct the genetic defect by transforming cells with remedial gene.
(d) Finally transplanting back these cells into the patient. In order to transfer remedial gene into patient's cell packaged retroviral method is employed. Retroviruses are modified in such a way that they no longer cause any disease.

10.6.2.2 Invivo gene therapy

It involves direct delivery of a remedial gene into the cells of a particular tissue of an effected individual. Isolation of cells from a patient is not required in this method. In gene construct, the remedial gene represents a sequence that encodes a protein that corrects the genetic defects of the patient. The remedial gene is under the control of tissue specific strong promoter. Some of the viral vectors such as Adenovirus retrovirus can be used to deliver remedial gene inside the patient.

10.6.2.3 Antisense therapy

Antisense therapy is another novel approach in gene therapy. Antisense approach involves the introduction of antisense gene that is complementary to all or a portion of a specific mRNA after its transcription in target cell. The transcription of sense and antisense gene in target cell produces sense and antisense mRNA. They hybridize and block the translation process. As a result the level of proteins or enzymes would be reduced drastically in a cell. Several genetic diseases are associated with defective protein production. In some human genetic diseases and cancers, genes can be either overexpressed or deregulated resulting in over production of proteins. Hence, antisense gene is constructed against hyper or hypo activity of these genes and marshall to a target cell to reduce the level of proteins. This antisense strategy would reduce the risk of disease to considerable extent.

Table 10.1. Clinical trials of diseases that are currently being considered for using somatic gene therapy

1. Cystic fibrosis	6. Atherosclerosis	11. Lung cancer
2. Colon cancer	7. Haemophilia	12. SCID
3. Breast cancer	8. Hypercholesteromia	13. Osteoporosis
4. AIDS	9. Leukaemia	14. Sickle cell anemia
5. Acute myeloid leukaemia	10. Liver cancer	15. Parkinson's disease

10.7 APPLICATIONS OF GENE THERAPY IN MEDICINE

(i) To target and destroy arterial clotting by using plasminogen activating factors against fibrin.

(ii) Curing Severe Combined Immuno Deficiency (SCID) by using Adenosine deaminase (ADA) gene.

(iii) Use of genetically modified and humanised antibodies against cell surface antigens to prevent the spread of breast cancer and prevent organ graft rejection.

(iv) Duchenne Muscular Dystrophy by myoblast transfer.

10.8 DNA FINGER PRINT

Since biotechnology has deeply inroaded in forensic science, there is no safe heavens for criminals. The suspected culprits can be brought to justice. Even paternal dispute, immigration problems can be effectively solved by a technique known as "DNA finger print". The technique was discovered by Alec Jeffreys of Leicester University in 1986.

Today, DNA finger print is potentially applicable in a wide range of scientific fields and will be described at the end of this topic.

The human haploid genome contains roughly three thousand million base pairs. Although DNA measures considerable length, it however, contributes only 10% of its coding for proteins and enzymes. In the remaining non coding sequence, there are areas of mini satellite repeats (Tandem repeats) scattered through the DNA. The exact number of tandem repeats varies significantly between individuals and the length of the mini satellite is different in different people. The region of tandem repeats in DNA is the main basis for DNA finger printing technique.

Methodology in DNA finger print:

Generally DNA finger print involves the following steps.

(i) Recovery of DNA

The main DNA source for finger print are blood, semen, hair or any biological sample etc. The DNA is extracted, pelleted and dissolved in suitable buffer for restriction digestion.

(ii) Restriction digestion and electrophoresis (RFLP)

The DNA sample is cut into their respective fragments. Restriction fragments of source DNA ensures differences in the length of tandem repeats. A restriction enzyme is selected in such way that it can cut on either side of the tandem repeat. Some of the examples of restriction enzymes used in Europe and USA for digestion are Hinf I, Hae III.

The minisatellite DNA fragments are then separated using agarose gel electrophoresis. This technique is generally called as Restriction Fragment Length Polymorphism (RFLP), as the DNA fragments exhibit polymorphism which will be unique to that sample. The normal size of minisatellite DNA fragments ranges between 2-25 kb. Generally λ phage DNA is used as marker DNA.

(iii) Southern blotting

When electrophoresis of sample is completed, DNA fragments are made single stranded and transferred from gel to nylon membrane by southern blotting or vacuum blotting. Single-locus probes and multi-locus probes are used in hybridisation process which is immediately done after southern blot. Single locus probes method is preferred in most of the criminal cases.

Once hybridization takes place between single stranded labelled probe DNA and complementary sequence of source DNA. It can be detected by subjecting to autoradiography. The position of minisatellite fragment can be detected by using labelled probe.

(iv) Autoradiography

In autoradiography in ^{32}P decays emitting β-particles which hit an X-ray film adjacent to the site of the bound probe. The time taken for autoradiography can go upto 10 days when small size of DNA sample is being analysed. If nucleic acid probe is not available, rapid chemiluminescent method of equivalent sensitivity is used (Fig. 10.6).

This relies on using chemically synthesised oligonucleotide probe.

(v) Match or no match

Most laboratories use a computerized video scanner to detect the bands and to compare them with the known control marker bands. Expertisation is required for the interpretation of DNA profiles of the disputed cases.

10.9 APPLICATIONS OF DNA FINGER PRINT

10.9.1 Forensic Science

DNA typing is used today in solving crimes or maternity cases as the technique provides valuable evidence to law enforcement authorities and court.

10.9.2 Medical

The application of DNA finger print profiling in the medical field includes cell population identification, tumour analysis, tissue culture cell line and paternity determination.

No Match

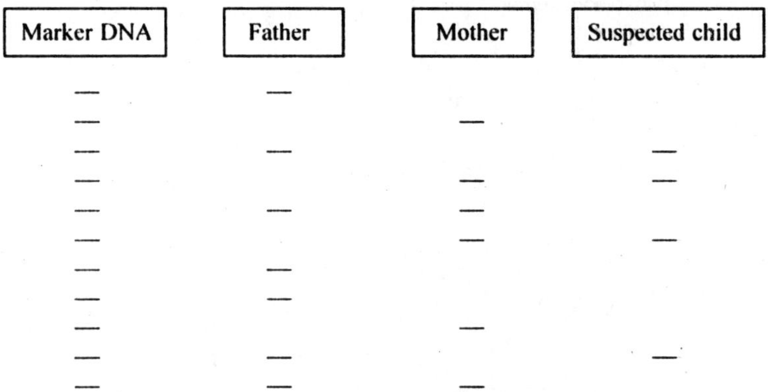

Match

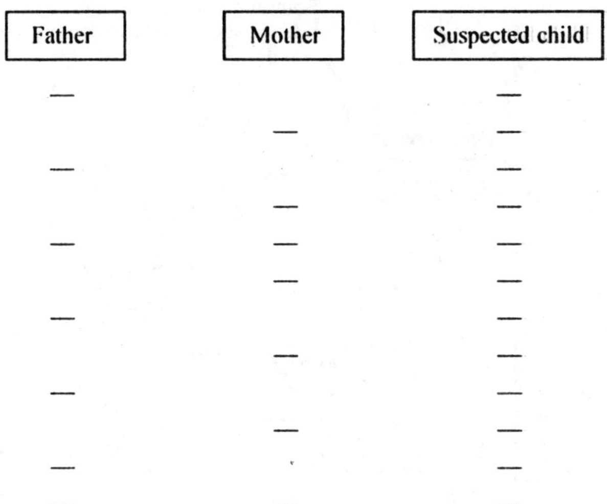

10.9.3 Wild Life

DNA profiling technique provides valuable information for better wild life management especially in breeding phenomenon in the population of tigers and elephants. It can even be used in wild life poaching.

10.9.4 Plant Science

DNA finger print helps in the detection of genetic diversity among varieties of economically important crops. RAPD analysis is used in profiling trait marker for use in breeding programme.

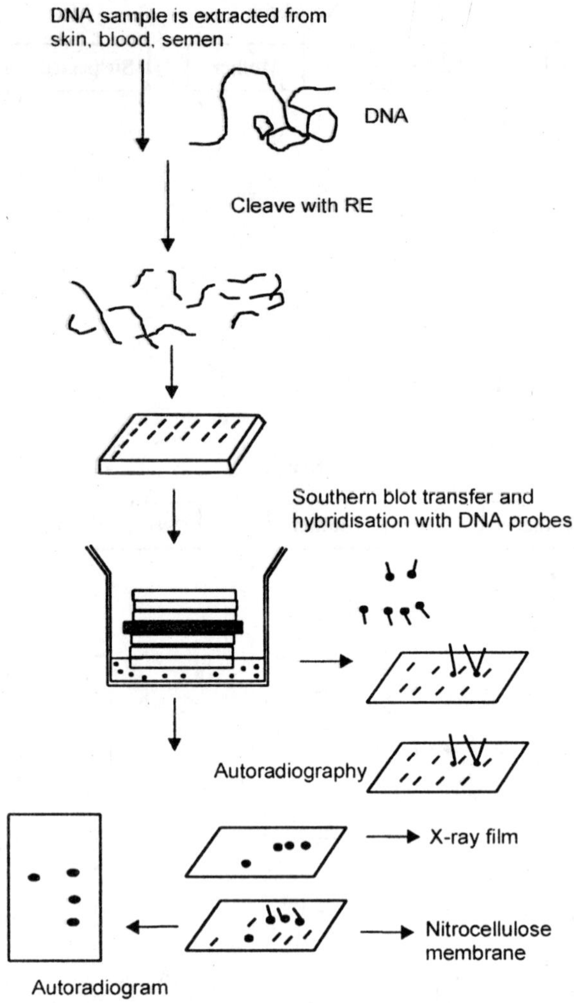

Fig. 10.6. General method of DNA finger print.

QUESTIONS

1. Give an account on the applications of Recombinant DNA Technology in medicine.

   (BU April 2001)
2. What are expression vectors? Describe methods of recombinant vaccine production.

   (BU Nov. 1999)
3. Explain how recombinant insulin is produced in *E. coli.*
4. Describe recombinant technique employed for the production of human growth hormone.

5. Describe biotechnological process used in the production of monoclonal antibodies.
6. Describe the method of producing monoclonal antibodies.

   (BU Nov. 1999)
7. "Gene therapy is a boon to mankind" – Discuss.

   (Nov. 1996) (BU May 1997)
8. What is DNA finger print. Add a note on its application.
9. What is RFLP. Explain its technique in the field of DNA finger print.
10. Discuss the role of Recombinant DNA technology for the production of new generation of drugs.

    (BU April 2002)

11

Biosafety Measures and Regulations for Recombinant DNA Work

Genetic manipulations have enabled to produce organisms capable of synthesising or modifying any number of useful proteins. The liability to transfer genes from one cell of a given species into another cell of a different species has significant impact in the field of rDNA technology. Genetic manipulation of DNA has already been made in microbial, plant and animal system. Certain countries have already permitted the use of some transgenic plants in the commercial agriculture. China was the first country to introduce in the early 1990s the viral-resistant tobacco and later tomato for commercial use. Subsequently in May 1994, Calgene Inc. USA introduced its "Flavr Savr" delayed ripening tomato for commercial use. By the end of May 1994 seven transgenic crops were approved for commercial use in USA. The recombinant DNA technology has further given boost to medical research with the development of important gene therapies.

Since recombinant DNA technology can potentially affect many aspects of modern society including agricultural production and medical treatment. There are significant ethical, legal, economic and social issues that need to be considered. For example, since its introduction in 1970's, serious doubts have been voiced by some individuals about the safety of recombinant DNA technology. It was felt that the ability to move genetic material from one organism to a different organism is sufficiently unique to warrant concern. At the time of first experiments in genetic engineering in USA, due to the entirely new nature of recombined species, they even decided to suspend to majority of the manipulation for a certain period of time. Fifteen years of experience have now demonstrated that these fears were totally baseless and also in vitro recombinations in microorganisms do not introduce any additional dangers to those observed with natural microorganisms.

Majority of industrial countries and international bodies concerned have decided to define five new criteria or risk to be considered in that manipulations on the basis of the model proposed by the office of Technology Assessment (OTA).

1. *Formation:* Deliberate or accidental creation of genetically modified microorganisms.
2. *Liberation:* Deliberate or accidental liberation of microorganisms at the work site or in the environment.
3. *Proliferation:* Multiplication, modifications and disappearance of these microorganisms in the environment and possible transfer of genetic material to other microorganisms.

11.1 BIOSAFETY METHODS (CONTAINMENTS)

The term biosafety methods or containments is used to describe the safe methods for managing infections agents in the laboratory environment. There are two types of containment one is biological containment other is physical containment.

11.1.1 Biological Containment

Biological containment are the vector like plasmids, viruses, for recombinant DNA and the host like bacterial, plant or animal cell. In the host vector is propagated in the laboratory. Any combination of vector and host which is to provide biological containment must be constructed to limit the infectivity of the vector to specific hosts.

11.1.2 Physical Containment

The objective of physical containment is to prevent the exposure of the researchers and the environment to the harmful agents. Physical containment is achieved through the use of following factors:

11.1.2.1 Laboratory Practice:

It involves strict adherence to standard microbiological practices, awareness of potential hazards, appropriate training for personnel, selection of safety practices, developing of adopting a biosafety.

11.1.2.2 Safety Equipment:

Safety equipment includes safety cabinets and a variety of enclosed containers. The biological safety cabinet is the principal device used to provide containment of infections aerosols generated by microbiological procedures. Safety equipment also includes items for protection such or gloves, coats, gowns, shoe covers, boots, and safety glasses etc.

11.1.2.3 Special Laboratory Design:

The design of the laboratory is important in providing a barrier to protect persons working in the facility but outside of the laboratory and those in the community from infections agents which man be accidentally released from the laboratory.

11.1.3 Biosafety Levels:

It consist of a combination of laboratory practices and techniques, safety equipment and laboratory facilities appropriate for the operations performed and the hazard posed by the infections agents. The guidelines for microbiological and biomedical laboratories suggest four biosafety levels depending on the nature and work. The proposed safety levels for work with recombinant DNA technique take into consideration the source of the donor DNA and its disease producing potential.

11.2 BIOSAFETY PRACTICAL

In basic laboratory guidelines are comprehensive and they are fundamental to all classes of laboratory. Some of the important values are as follows.

1. Mouth pipetting is prohibited.
2. Eating, drinking, storing food and applying cosmetics is generally prohibited in the laboratory.
3. Work surfaces should be decontaminated at least once a day after any spill of potentially dangerous material.
4. Working persons should wash their hands after handling infectious materials and animals, when leaving the laboratory.
5. All technical procedure should be performed in a way that minimizes the creation of a aerosol.
6. All contaminated liquid or solid materials should be decontaminated before disposal or reuse. Contaminated materials that are to be autoclaved should be placed in durable leakproof container.
7. Laboratory coats, or uniforms should be worn in the laboratory. Contaminated clothing should be disinfected by appropriate means.
8. Safety glasses, face shield or other protective devices should be worm when necessary to protect the eyes and face from splashes and impacting objects.
9. The use of hypodermic needles and syringes should be restricted to parental infections and aspirations of fluids from laboratory animals and diaphragm vaccines bottles.
10. All spills, accidents and overt or potential exposures to infectious materials should be reported to concern authorities.
11. Baseline serum samples may be collected from and training in laboratory safety is provided. A safety or operations manual that identifies known and potential hazards and that specifics practices

and procedures to minimise or eliminate such practices working personnel should be advised of special hazards and required to follow standard practices and procedures.

11.3 LABORATORY DESIGN

1. Ample space should be presence forth safe conduct of laboratory procedure.
2. Walls, ceiling, and floors, should be smooth, easily cleanable, impermeable to liquids.
3. Adequate illumination should be ensured for carrying out all activities. Undesirable reflections is to be avoided.
4. Spaced facilities should be provided for the safe handling and storage of solvents, radioactive materials.
5. Safety systems should cover five, electrical emergencies, emergency shower, and eyewash facilities.
6. A good quality and dependable water supply is essential. The public water system must be protected by a back-flow preventer.
7. There should be energy lighting to permit safe exit. A stand by generator with automatic cut-off is desirable for the support of essential equipment-incubators, freezers etc. in particular, it is indispensable for the ventilation of animal cages.

11.4 LABORATORY EQUIPMENT

The risk of an infection can be minimized by the use of safety laboratory equipment, practices and facilities. In selecting safe laboratory equipment, the general principle should be considered include.

(1) Designed to limit or prevent contact between the operators and the infections agent.
(2) Equipments are corrosion-resistant and meet structural strength requirement.
(3) Fabrigated to be free of burrs and shard edges.
(4) Biological safety cabinets to be used whenever procedure with a high potential for creating hazardous aerosols are conducted. This may include centrifugations, blending, sonic disruption, opening containers of infectious materials, intranasal inoculation of animals, and harvesting infected tissues from animals or eggs.
(5) Pipetting aids-to replace mouth pipetting.
(6) Loop micro incinerators—to reduce aerosol production.
(7) Autoclave to sterilize contaminated material.

11.5 EMERGENCY PROCEDURES

Emergency plans should include.

(1) Breakage and spillage.

(2) Accidental injection, cuts and abrasions.
(3) Accidental injection of potentially hazardous material.
(4) A potentially hazardous aerosol release.
(5) Emergency services-fire, flood and natural disaster.
(6) Breakage of tubes in centrifuges not having safety cups.

11.6 FUNCTIONS OF ADVISORY COMMITTEE AND STATUTORY BODY

11.6.1 Recombinant DNA Advisory Committee (RDAC)

This committee take note of developments at national and international levels in biotechnology towards the current safety regulations for India on recombinant research use and applications. This committee meets in 6 months. The specific terms of reference for RDAC include. To evolve long term policy for research and development in Recombinant DNA research. To recommended type of training programme for technicians and research follows for making them adequately aware of hazards.

11.6.2 Implementation Committees: (Institutional Biosafety Committee (IBSC))

Institutional Biosafety Committee are to be constituted in all centres engaged in genetic engineering research. This committee will constitute:

(i) head of the institution (ii) 3 or more scientists engaged in DNA work or molecular biology with an outside expert in the relevant discipline. (iii) A member with medical qualifications-Biosafety officer. (iv) One member nominated DBT (Dept. of Biotechnology) Govt. of India.

11.6.3 Review Committee on Genetic Manipulation (RCGM)

The RCGM will have the following composition.

(i) Indian Council of Medical Research (ICMR).
(ii) Indian Council of Agricultural Research (ICAR).
(iii) Council of Scientific and Industrial Research (CSIR).
(iv) Three experts in individual capacity.
(v) Department of Science and Technology (DST).

The RCGM will have the functions like establish procedural guidance manual-procedure for regulatory process with respect to activity involving genetically engineered organisms in research, production and applications related to environmental safety. To review the reports in all approved ongoing research projects involving high risk category and control of field experiments. to advice customs authorities on import of biologically active material, genetically engineered substances or products. To assist Department of Industrial developments, Banks towards clearance of applications in setting up industrial based on genetically engineered organisms. To assist the Bureau of Indian standards to evolve standards for biologics produced by rDNA technology. To advice on intellectual property rights with respect to rDNA technology on patients.

11.6.4 Genetic Engineering Approval Committee (GEAC)

Genetic Engineering Approval Committee will function under the Department of Environment as statutory body for review and approval of activities involving larger scale use of genetically engineered organisms and their products in research and development environmental release and field applications. The important functions of committee include:

(i) Import, export, manufacture, selling of any microorganisms or genetically engineered substances or cells including food stuff and additing that contains products derived by Gene Therapy.
(ii) Discharge of genetically engineered, hospitals and related areas into environment.
(iii) Large scale use of genetically engineered organisms from laboratory and hospitals.
(iv) Deliberate release of genetically engineered organisms.

The composition of the committee would include:

1. Chairman—Additional Secretary, Dept. of Environment.
2. Representatives of Concerned Agencies and Departments like—Dept. of Biotechnology, ACST, Dept. of ocean development.
3. Expert members.
   (i) Director General - ICAR
   (ii) Director General - ICMR
   (iii) Director General - CSIR
   (iv) Director General - Health Services
   (v) Plant protection adviser
   (vi) Chairman Pollution Control Board.

QUESTIONS

1. Discuss nature of Biosafety methods and guidelines for recombinant DNA work.
2. Explain the role of different advisory committees in controlling rDNA work.

SHORT QUESTIONS

1. Biosafety Methods (BU Oct. 2001)
2. Genetic Engineering Approval Committee
3. Containment Levels
4. Laboratory Design

Part B
Environmental Biotechnology

1

Renewable and Non-Renewable Resources

Natural resources are the richest treasure on this green planet. This highly diversified resource has offered immense value in the development of greater quality of life and society. These are the ultimate lifeline for human development and his survival. Energy, matter and diversity are combinedly called as natural resources. All life forms are completely dependent on these natural resources. Hence, there is a greater degree of interaction existing between different ecosystems. Natural resources are part of our atmosphere, hydrosphere and lithosphere. The biosphere was evolved due to prolonged and co-ordinated interaction between different ecospheres. The components of bio-sphere, atmosphere and hydrosphere are perfectly involved in supporting life.

Natural resources are of two types, inexhaustible and exhaustible. Some resources are available to man for millions of years and are said to be inexhaustible e.g., water, forest, life forms. Some resources are fixed and exhaustible after continuous use, e.g., coal and petrol. These deposits on the planet are finite.

More precisely, natural resources can be classified basically into two forms. (a) Renewable and (b) non-renewable. These two are part of the natural systems and have tremendous impact on human welfare. The progressive formation, development and accumulation of the resources are the integral component of nature. India has today among the world's largest programme for renewable energy. During last two decades, several renewable energy technology have been developed and deployed in villages and cities. There are path breaking success of various renewable energy protect in areas like Laddak, Sunderbans, Chattisgarh etc.

1.1 RENEWABLE RESOURCES

Renewable resources are purely conventional types. Classical examples are water, forest, food crops, plant species, wild life, livestock and aquatic life. These resources can be called as "replacable resources". Forest resources, food resources, aquatic, wild life and livestock resources are prominent renewable source of energy to sustain life on the earth. These resources can be replaced from time to time due to their life cycle. The food resource can be harvested at regular period of time and are cycled for indefinite period of time. Some resources do not have life but can be recycled (water resource) and are renewable.

1.1.1 Water Resource

Water is indispensable for every life activity of living organisms. It is required for irrigation, industries, power generation and importantly for domestic use. 97% of the water available on earth is salt water and the rest is fresh water. Agriculture sector consumes largest amount of water. The ground water table is depleting due to excessive usage. An average of 2.3 litres of water is required by a person in his daily life.

1.1.2 Forest as Renewable Resource

Biomass can provide several conventional renewable energy source. The agricultural waste, forest wood is used as biomass fuel and they are renewable. One of the important renewable energy source is forest. It constitutes of more than 90% of global biomass. Some of the richest and hot spot bio diversities are in forest ecosystem. Forest ecosystem is an integral part of the biosphere dominated by tree species (Table 1.1).

Forest contains rich diversity of fibres, fuelwood, fodder, timber yielding trees and a variety of medicinal plants. Many wood industries are dependent on forest for wood and other raw materials. Forest also has a rich treasure of resin and gum, fibre and latex. Forest regulates climatic conditions especially rain. They trap solar energy and transform into biomass.

Indian subcontinent contains some important biospheres. The western ghats, which runs nearly 1200 km from Gujarat to the end of Kerala is considered as one of the richest hot spot biodiversity in the world. The dominated species are mainly tree canopies, shrubs, epiphytes and medicinal plants. They provide high durability wood including teak and rosewood for commercial use. Based on the types of vegetation they are classified as evergreen forest, semi evergreen forest, deciduous forest and scrubby jungles.

The forest soils are rich in useful microorganisms. The accumulated waste biomass undergoes biodegradation process by soil microorganisms and are converted into humus mass, rich in nutrients. They can increase the fertility of the soil.

Table 1.1 Some important biospheres of India

Biosphere	State
Nilgiri	Tamil Nadu
Mannar	Tamil Nadu and Kerala
Kanha	Madhya Pradesh
Kaziranga	Assam
Sunderbans	West Bengal
Cherapunji	Meghalaya

1.1.3 Agriculture as Renewable Resource

Agriculture is typical example for renewable source of energy. Agriculture management not only involves crop culture but also focuses on fertile soil, water, pest control, fertiliser applications etc. Annual cultivation of two to three types of crops are now possible on the same land. The practice of crop rotation yields good result by increasing soil fertility. A good agriculture system can provide millions of tonnes of food grains. The world food production is estimated to be more than 200 million tonnes according to agri. reports. Apart from food being a renewable source of energy, agricultural waste can be an excellent raw material for gas, alcohol and manure production. Modern agriculture practice created several environmental problems i.e., increased use of pesticides, herbicides and accumation of heavy metals.

1.1.4 Livestock as Renewable Source

Domestic animals are important resources. They provide milk, meat, wool and other products. The most common livestock are cattle, sheep, goat and cow. The dung produced by excretory activity of livestock is an excellent raw material for bio gas and manure production.

This ecofriendly biomass can meet major chunk of the energy demand especially from rural sector. Hence, the biomass produced by livestock has potential environmental impact.

1.1.5 Major Consumer Items as Renewable Source

Man realises useful resources from nature. Several centuries ago he discovered that plant could be used as a source for food and fibre. These are the major consumer items that are extensively exploited, but fortunately they are renewable sources. Food and fibres are the very basic need for human life. Fibres are perhaps the most sustainable material for human civilisation. Though the dependency on these two sources has increased due to population growth, continuous replacement of renewable sources can over ride the depletion of natural resources.

1.1.5.1 *Food*: Food is a complex array of chemicals. It consists of carbohydrates, proteins and lipids. The entire scenario of food chain shows that all living organisms including man are dependent on plant for food. The entire energy consumed during the life activity are derived from food. The main energy flow in biosphere comes from the assimilation of food material in living organisms. Although several thousands of plant species are available, only a handful of plants are accountable as major food producer on this planet.

Cereals, millets and tubers can provide major staple food. Starch, cellulose and hemicelluloses are enriched with the fuel molecule, glucose. Plants have been classified into two major divisions—monocotyledons and dicotyledonous. Among dicotyledons plants, legumes are the second most important group of food, which belongs to the family Leguminosae (Table 1.2). They are essential part of our diet and are commonly known as pulses. The nuts possess good amount of proteins, adequate lipids and low water content. The common pulses are soyabean, grams, lentils and peas. Soyabean among dicotyledons contain the richest source of proteins. India has the distinction of being the world's largest producer of pulses and their seeds are rich in proteins, carbohydrates, fat, vitamins and minerals.

Cereals and millets belong to the family Graminae. They are well adapted to grow in all agroclimatic conditions and are cultivated throughout the world. India, China, USA, Indonesia, etc. are the largest producers of cereals and millets. Cereals include wheat, rice, maize, barley and millets are small grains from plants like sorghum and ragi.

Table 1.2. Classification of foods

Major cereals	*legumes*
Wheat	Black gram, moth bean
Rice	green gram, chick pea,
Maize	soyabean, kidney bean, Ground nuts cluster bean, cow pea, garden pea and pigeon pea
Minor cereals	
Oat	Barley
Nuts	almonds,
Ragi	cashewnut and wallnut
Millets	*Vegetables*
Sorghum Pearl millet Finger millet	Raddish, carrot, turnip, sweet potato, cabbage, garlic, onion, tomato, cauliflower, spinach, lady's finger etc.

1.1.5.2 *Fibres*: Fibre is the second largest major consumer item after food. Human beings are dependent on plants for fibre products. Cloth is the most indispensable prerequisite for human beings. Cloths are being made out of natural fibres obtained from several plant species. Two thousand different species of plants can provide fibres. Man realises the significance and utility of fibre as a clothing material. Fibres are elongated cells and occurs in different parts of the plant like root, stem, leaf and fruits. According to their origin they are categorised into surface fibre, wood fibres, bast fibres and leaf fibres. Surface fibres are hair like out growth on the seeds of some plants. Wood fibres are obtained from wood. Commercially bast fibres are durable and important and are obtained from phloem fibres or cortical fibres. Leaf fibres are sclerenchyma cells associated with the vascular bundle strands. Plant fibres are commercially significant due to their tensile strength and chemical composition. They are further classified based on their commercial uses. They are textile fibres from which various fabrics, cordges, netting material can be prepared. Brush fibres are tough and used for making brushes and brooms. Filling fibres are essential for stuffing and filling while making beds and furniture (Table 1.3).

Table 1.3. Major fibre items

Textile fibres	Brush fibres	Filling fibres	Vegetable fibres
Cotton	Broom corn	Red silk	Luffa
Flax	(Sorghum)	Cotton	
Hemp	Toddy palm	White silk cotton	Sponges
Ramie	Coconut	Madar	
Rosette			
Agavefibre			
Oak			
Pineapple			

Our forest ecosystem has rich diversity of fibre yielding plants. Various wood fibres are used for making papers. In view of commercial significance of fibres, man has continued to exploit fibre products. Mass cultivation of fibre plants can meet the challenge posed by dwindling reserves. Some of the fibre plants like Agave, Bromliads and flour are well adapted to grow on dry sterile soil. They can be grown in wastelands and on either sides of railway track, to prevent soil erosion. This would help in binding of soil and increase soil fertility.

1.2 NON-RENEWABLE RESOURCES

Non-renewable resources cannot be replaced. Since they are nonlife forms, regeration or cyclic process is not possible. In contrast to renewable resources, non-renewable resources do not have life cycle. Examples of non-renewable resources are minerals soil reserves, fossil fuels and other land resources.

Minerals: Mineral resources mainly include inorganic and organic substances. Number of minerals such as copper, iron, coal, petrol, etc. are essential for human beings as they are commercially used in industry. They are used to make iron implements, machines and in petroleum industry. Due to continuous excavation of mineral resources their depletion in taking place at an alarming rate. The world oil deposits are concentrated around Persian gulf. More than 60% of the world's oil belt are around south west Asian countries. Since oil is a non-renewable resource greater concern has been shown to look for other substitutes.

1.2.1 Fuels

Fuels are non-renewable source of energy. They cannot be replaced once their deposits are exhausted. Petroleum is an example for fossil fuel. The origin of petroleum can be traced back to plants and animals that lived on terrestrial part of earth and in the sea millions of years ago; these dead organisms inside the earth and sea underwent degradation process due to bacterial action. It is also coupled with the action of earth heat and pressure. This biogeochemical process was carried out over millions of years in the earth's crust and converted to liquid hydrocarbons, namely petroleum. Besides, the gaseous hydroxides were also produced of a product of earth's activity, that are known as natural gases. Fossil fuel petroleum is invariably required as an automobile fuel. The crude petroleum obtained by distillation is called gasoline, a source of energy in the automobile internal combustion engine.

Just after the world war, most of the European nations started using oil. The exact use of petroleum as a source of energy began by 1900. This liquid fuel showed several advantages like speed, refuelling at sea and was economically beneficial to the sailors.

1.2.2 Soil

Soil is formed through series of weathering and decomposition processes. It takes 50 years to form one layer of top soil. Soil can support plant life and in turn animal life. It can store water efficiently. Various types of soils such as red soil, sandy soil and black soil support different types of vegetation. Loss of soil fertility is a serious problem. Loss of soil texture is due to erosion and cannot be replaced. Deforestation, adverse climatic effects, poor agricultural management are some of the reasons for soil erosion. It can pose serious environmental impact in view of loss of fertility due to pesticide accumulation and soil erosion. Hence, soil management is the top most priority today to check erosion and maintain fertility.

1.2.2.1 *Soil conservation:* Soil *conservation* can be managed by good agronomic practice, dry farming practice. Some of the agronomic practices are contour farming, tillage and crop rotation i.e., sowing of leguminous crop and mixed cropping. Rotation of cereals and legumes is useful in maintaining production of fertility of soil. The use of legumes include nitrogen content in the soil. The nitrifying bacteria present in root nodules of legume plants fixes atmospheric nitrogen into nitrate, nitrite and finally ammonium salts. Due to death and decay of leguminous plants substantial amount of nitrogen compounds are retained in the soil by decomposition. Mulching of different kind can check soil erosion, increase soil fertility and minimise moisture evaporation. The cultivation of grasses in wastelands can not only provide protection to soil but can improve soil fertility by binding soil particles. They can even serve as a good fodder source for livestock. Aforestation is an affordable programme to balance ecosystem. Plantation of several fast growing plants prevents wind erosion of top soil. The leaves and branches of trees, shrubs, grasses helps to break the force rain water.

QUESTIONS

1. Mention "renewable", "non-renewable" and non-conventional source of energy. Highlight the need to find new renewable and non-conventional source of energy.

   (BU Nov. 1997)
2. Define renewable and non-renewable source of energy. Highlight their significance in view of future energy crisis.
3. Explain major consumer items as Renewable source.

SHORT QUESTIONS

1. Non-renewable source
2. Renewable source
3. Energy plantations

2

Conventional Fuels and Environmental Impact

All conventional fuels are renewable source of energy. Due to population growth, the pressure on demand for all types of fuels has increased. Major fuel consumption takes place in industrial, urbanised and domestic sectors. Industrialised nations are major consumers of many conventional fuels. US is the largest consumer of fossil fuel. Overall fossil fuel reserve is depleting at an alarming rate.

Oil, coal and natural gas represents the concentrated energy source of the earth. These are the accumulated residue of billions of years of photosynthetic activity. The over usage of coal, firewood and petroleum oil has a serious negative environmental impact. Combustion of fossil fuel increases CO_2 in the atmosphere enhancing the possibilities of global warming i.e. green house effect.

2.1 FIRE WOOD

Fuel wood is used for heating and cooking. Nearly one third of the world population uses fuel wood. The wood is a renewable source. However, it is limited. More intelligent use of wood resources will not only enhance the supply but also conserve the resource. Forest, scrubby jungles are the chief source for fire wood (Table 2.1). The present annual contribution of fuel wood in the United States is approximately 7.5 quads. Potential uses of firewood cover wide spectrum of industrial, agricultural and residential sectors. Transportation of wood is not economical. Wood is made into pellets, chips or saw dust are appropriate for automatic stocking methods and will improve the convenience of wood energy. In order to capture complete energy from wood burning process, designing of wood burning appliances are essentially required. There are three areas of concern in the design of wood burning process. Safety, efficiency and pollution. If the problems can be tackled efficiently, perhaps wood energy would be considered as an effective alternative to fossil fuel.

2.1.1 Environmental Impact

Technology has helped us to increase the energy efficiency of wood burning devices. The most significant improvement is the ability to construct more air tight appliances that can control the inflow of combustion of air. Presently, effort has been made to increase combustion efficiency of firewood. Different types of wood burning equipments are now available for domestic purpose especially in rural areas. These include traditional fire place, air tight stoves and wood burning furnaces. Sometime back the Bangalore based research institute ASTRA (Applications of Science and Technology for Rural Areas) developed efficient wood burning appliances for rural sector. The heating efficiency and less pollution are some of the special features of ASTRA appliances.

Energy from wood based fuels undoubtedly have positive impact on present energy crisis. Wood combustion does cause pollution. Air borne emission by burning process contains potentially hazardous chemicals. The main pollutants are carbon dioxide and carbon monoxide which impair respiratory efficiency of the lungs. Substantial amount of sulphur and nitrogen oxides are released into the environment.

2.2 ENERGY PLANTATION AND SOCIAL FORESTRY

In the near future, as petrol and nuclear fuel would increase in cost, it is predicted that firewood forms will be used extensively. Hence, pressure will be more on firewood sources. Already deforestation is taking place in the vicinity of rural areas and nearby forest. Social forestry is an effective programme to combat deforestation problem. Energy plantation can reduce pressure on future firewood demand.

Energy plantation is simply growing trees for fuel availability. Energy plantation is renewable wood fuel resources. Currently, plantation programme is given prominence by the Government. A greater enthusiasm towards energy plantation results in various parts like soil forestry, silviculture and horticulture. The depleting forest resources can be strengthened by social forestry, wasteland management, roadside land, either sides of canals and railway lines are some of the barren land which have been used for social forestry. Some of the fast growing, dryland adapted plants are *Albizia, Azadiracta indica, Butea monosperma, Acacia, Eucalyptus* sp., *Madhuca Indica.*

Active participation of several NGO's (Non Governmental organisations) in social forestry have taken significant momentum. Growing our own fuel wood is not only economically beneficial but also strengthen energy management. Hence, man's first fuel is once again the most acceptable fuel.

Table 2.1. Fuel wood for different regions

Tropical regions	Arid and Semi arid regions	Tropical high lands
Muntingia calabura	*Acacia* sp.	*Alnus rubra*
Calliandra sp.	*Albizia lebbeck*	*Eucalyptus grandia*
Casuarina sp.	*Cassia* sp.	*Acacia* sp.
Acacia auricaliformis	*Emblica officinalis* *Parkinsonia aculata*	*Grevillea robusta*
Gmelia arorea	*Eucalyptus* sp.	
Terminalia	*Azadiracta indica* *Zizyphus mauritiana*	

2.3 PLANT AND ANIMAL WASTE

Plant and animal waste as a whole are renewable source. In the recent years there have been quantum shift in the utilisation of waste reserve. Forest, agricultural field, plantation forms, poultry, dairy industries and residential areas are some of the places, where wastes are generated. Animal waste is generated mainly as excreta.

2.3.1 Plant Waste

In agriculture field considerable amount of biomass waste in generated by wheat and paddy straw in addition to sugarcane trash, maize stalk, sorghum stalk, etc. Many agrobased industries produced fruit waste residue after the processing are mango, pineapple and banana peels. Residential sites, snack bars also contribute to considerable amount of waste.

2.3.2 Animal Waste

Every year thousands of tonnes of animal waste is generated. Animal wastes are cow dung, poultry's animal excreta, carcass, cattle urine, heads and bones of fishes, prawn waste, frog etc., constitute marine waste.

In India, annual output of livestock waste is accountable for more than 75% of the total excreta and farm yard. Waste from livestock excreta contains 34×10^6 N and 13×10^6 P_2O_5. This can be used as broad organic manure in conjunction with chemical fertiliser. Cow dung in India is extensively used as a manure as well as bio fuel. According to the studies one tonne of cow dung manure contains 5 kg N, 2-5 Kg P_2O_5 and 5 kg potash, besides plenty of macro and micro nutrients. Cow dung also has antifungal, antibacterial and insecticidal properties. Cow dung as a renewable source of energy contribute nitrogen, phosphorus and potash and helps in enrichment of fertility of the soil.

Several plant based industries like coconut coir, cashewnut, sugar factories, tea factories, rice mills, cotton and jute industries are the main attractive places for waste byproducts. The sugarcane bagasse, molasses are substrates for ethanol production. Coconut shells, coir and paddy husks are used as conventional fuel. The overview of waste from plant and animals are summarised in the Table 2.2. The mixture of wood and bark waste is used as fuel known as "hogfuel". The burning of hog fuel generates high stream of pressure helps in drying waste. The Government of India encourages recycling of waste for the production of many unconventional fuels. Some of the government bodies like the Department of Science and Technology (DST), Department of Biotechnology (DBT), have launched several research projects on waste management and utilisation. DST has implemented many income generating waste management programme for rural people especially in the areas like composting and vermi composting.

Table 2.2. Various plant and animal wastes and byproducts

Waste/Byproducts		
Crop residue/forestry	Agro Industry	Dairy, poultry and aquaculture
Paddy, wheat straw, sorghum stalk, groundnut shell, sugarcane trash, cotton stalk, jute sticks, coconut shell and coir, tea and coffee residue, wood and bark residue, wet and dried leaves	Mango, banana peels, rice husk, rice, bran, cotton dust, castor seed cake, jute mill waste, groundnut cake, sugarcane bagasse, molasses and sawdust.	Cow dung, sheep and goat waste, poultry waste, prawn and frog waste.

2.4 GAS

Natural gas is hydrocarbon. They were produced in earth crust and sea. The dead plants and animals underlying the earth and sea were subjected to microbial action and severe earth's heat and pressure for millions of years ago and were transformed into fossil fuel like petroleum. Gaseous hydrocarbons constitute natural gas. Natural gas is the most common source for methanol production and as automobile fuel. Methanol energy content is equal to one half of gasoline. Most of the atmospheric methane is produced by anaerobic bacteria in swamps and bogs, smaller percentage is released by natural gas wells.

Natural gas or oil may be present together or separately. If it is together, the gas may be in top as gas cap or dissolved in oil. This is known as associated gas. Gas may be present freely without association with oil. From the natural gas certain hydrocarbons can be obtained in liquid form known as Natural Gas Liquid (NGL). These are valuable energy resource in the world. NGL compounds are propane, butane, pentane, hexane and heptane.

India contains several natural gas reserves found in Andhra Pradesh, Gujarat, Rajasthan, Tamilnadu and eastern off shore coastal areas. The Oil and Natural Gas Commission (ONGC) India procures 3-8 million cubic litres of gas fron. gas reserve.

2.4.1 Compressed Natural Gas (CNG)

The Supreme Court of India passed an order in 2001 and directed the Delhi administration to convert all the public transport buses into CNG. The order was passed to minimise air pollution in Delhi. CNG or compressed natural gas is a mixture of hydrocarbon (90%) and, methane in gaseous forms. Due to its low energy density, it is compressed to a pressure of 200 to 250 kg/ cm^2. It is colourless, non-toxic, lighter than air and inflammable. All vehicles including diesel engine can be converted into CNG. It is safest, ecofriendly gaseous fuel. It has higher octane number and lower cectane number makes it superior than petrol. CNG does not contain lead or benzene. Hence lead and benzene pollution is eliminated. Use of CNG reduces carbon monoxide emission by 70-90%. Even ground water pollution can be averted since it is stored in cylinders.

2.5 COAL

Coal is highly condensed carbon and an important fossil fuel. It is formed due to carbonisation of plant. Carbohydrate in coal was formed from the debris (organic matter) of giant tree ferns and other vegetation that was growing in swamps and other areas. Some three hundred million years ago, when the gigantic tree ferns and other plants died, they were under sediments that prevented their aerial oxidation. This in association with earth's high pressure resulted in their conversion to coal by carbonisation. Carbonated coal when heated gives rise to a variety of important organic compounds. It also contains burried variety of inorganic ions like Na, K, Ca, Al, etc. all these finally turn to ash residue. Formation of coal involves series of transformations. Initially organic matter is metamorphosed. Peat is formed as brown and porous material. It contains plant remains with high amount of moisture. The weight of overlying rock coupled with the earth's heat drive out moisture from the burried peat and metamorphosed into lignite. These lignites are transformed into bituminous coal. Chemically plant carbohydrate is converted to hydrocarbons and then into condensed carbon. About 600 billion tonnes of coal was estimated to be present below the earth. Coal can be subjected for gassification and liquification. The coal is converted to carbon monoxide and hydrogen which is hydrogenated at 200°C in presence of iron as a catalyst to get a mixture of hydrocarbon.

2.5.1 Environmental Impact

Coal is invariably used in some industries and in domestic use. Over many centuries coal was commonly used as a driving energy force in railway engines. However, diesel train gradually replaced coal engines in railway system. Coal is extensively used for cooking and for power generation sector. The thermal stations burning coal, release ash as fine dust particle. The particulate emission due to coal burning is accumulated in air. The particles are sulphur dioxide, nitrogen oxide, carbon monoxide and hydrocarbon. The sulphurdioxide can mix with water moisture of air and results in the formation of sulphuric acid mist. The acid is the most condensed unburnt carbon particles formed as acid smut. The combustion of coal also releases oxides of nitrogen. Carbon dioxide pollution by coal burning depends upon its carbon content. Thus fossil fuel (coal) causes many negative environmental implications. The gaseous pollutants like CO_2, CO, may pose serious health implications. However, problems are confined mainly to urban sector due to their automobile density.

QUESTIONS

1. What are conventional fuels? Explain the types and environmental impact of conventional fuels.
2. Discuss merits and demerits of conventional fuels.

SHORT QUESTIONS

1. Coal
2. Compressed Natural Gas (CNG)
3. Fire wood
4. Conventional fuel and their impact

(BU Nov. 1999)

3

Non Conventional Fuels and Environmental Impact

In recent years, several attempts have been made to exploit natural resources to tap bioenergy. The effective utilisation of microorganisms and waste biomass for the production of bioenergy are some of the land marks in biotechnology. The contribution of Methanogenic bacteria in waste management enables us to solve environmental problems.

The gradual depletion of fossil fuels, future energy crisis and over exploitation of bioresources have made man to resident himself and search for alternative energy resources. Nature is a rich treasure of energy. The solar energy is efficiently trapped and stored in plants and in turn energy flow is directed to animal system. In order to overcome overall energy crisis a comprehensive studies has already been made in the direction of using non conventional fuels. Some of the waste like kitchen refuse, agriculture waste and municipal waste are the real treasure of energy source. The fuel molecule glucose is locked in cellulose and semi cellulose as a major component of plant waste. Ability to produce ethanol and other alcohol by microbial biodegradation of waste material is being considered as one of the remarkable achievements in microbial biotechnology. The Brazil's nationwide alcohol programme can be considered as a new dimension in exploiting alcohol as automobile fuel. Besides, microbial hydrogen production as a potential fuel and tapping plants for petroleum are some of the preset and future strategies that have already been made in scientific sector. Man's rethink on non-conventional fuel is a challenging task before the ever growing population.

3.1 METHANOGENIC BACTERIA AND BIOGAS

Methanotrophs are a group of microbes which utilise carbon dioxide as a electron acceptor and methane is produced during biomethanation. These are gram positive in nature. The cell walls structure contains pseudomurein as a key component. Biomethanation or Biomethanogenesis is an ancient process of conversion of biomass into energy by Methanobacteria like *Methanococcus* and *Methanosarcina.* All the three genera of methane bacteria have the ability to obtain energy from the oxidation of hydrogen gas and reduction of carbon dioxide.

$$4H_2 + CO_2 \rightarrow CH_4 + 2H_2O$$

In 1976, a person by name volta, discovered the presence of methane in the biomethanogenesis and called it as "*marsh gas*" or "*gober gas*". The methanogenic bacteria reduce carbon dioxide to methane and consume hydrogen. The methanogenic bacteria are highly sensitive to oxygen and thrive well in presence of hydrogen and carbon dioxide. *Methanobacterium formicicum and methanospirillum hungati* produces methane derived from aromatic compounds under strict anaerobic conditions. The methanogenic bacteria are a diverse group of anaerobes and are thermophilic. The methanogenic bacteria produce methane as the last product of a complex series of reaction.

3.1.1 Biogas

Biogas is a bio fuel, produces by microbial action on organic matter. Biogas production involves complex mixture of anaerobic bacteria which convert up to 90% of the organic matter to methane and carbon dioxide. This gas contains nearly 80% of methane, 15-30% carbon dioxide, besides hydrogen, nitrogen and traces of hydrogen sulphide. Biogas is lighter than air and has a calorific value of 8600 kcal/m^3. The presence of carbon dioxide in biogas reduces or it lowers the calorific value of the biogas. The level of CO_2 can be reduced by various techniques. Even traces of hydrogen sulphide can be removed for commercial use.

3.1.1.1 *Biogas production*: Generally, biogass production involves three important stages (i) Solubilisation and hydrolysis of organic compounds (ii) acidogenesis (iii) Methanogenesis

3.1.1.2 *Waste Material for Biogas production:* Both rural and urban areas generate ample quantity of biowaste. They are cow dung, plant waste, agricultural waste, kitchen garbage and sewage waste (Table 3.1).

3.1.1.3 *Hydrolysis:* In the first stage of enzymatic hydrolysis, the complex polymers are hydrolysed into organic acids and alcohol. Several hydrolytic bacteria have been used. Many soluble substance, after hydrolysis acts as substrate for microorganisms.

3.1.1.4 *Acidogenesis*: In the second stage many soluble organic compounds are converted to organic acids. There are three important groups of bacteria used in the second stage. First group convert complex organic substance into butyric and lactic acid. Second group, acetogenic bacteria converts organic acids into lactic acids by hydrogen and carbon dioxide. The third group are the methanogenic bacteria which serve as substrate for the final stage of decomposition.

Biogas production is carried out in water tight cylinder with a side opening as inlet. Depending upon the nature and amount of gas production different types of biogas plants can be constructed.

Table 3.1. Raw materials for Biogas production

Animal waste	Plant waste	Garbage
Cow dung, poultry waste	Leaves, grasses, weeds, aquatic weeds, grasses, Sugarcane bagasse, paper waste	Vegetable peel, Fruits peel, sewage waste.

3.1.1.4 *Construction of Biogas plant:* Simple biogas plant used in the countryside, generally comprises of four parts. The inlet pipe, the outlet chamber, the digestor fermentation and gas collection.

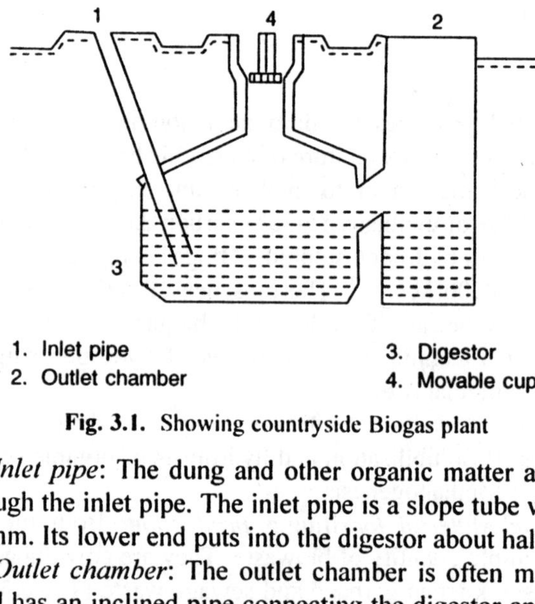

Fig. 3.1. Showing countryside Biogas plant

3.1.1.4.1 *Inlet pipe*: The dung and other organic matter are put into the digestor through the inlet pipe. The inlet pipe is a slope tube with a diameter around 300 mm. Its lower end puts into the digestor about half of the height.

3.1.1.4.2 *Outlet chamber*: The outlet chamber is often made in vertical well type and has an inclined pipe connecting the digestor and is positioned at 180° to the inlet pit to avoid the raw material dropping into the outlet chamber before fermentation. The lower end of the inclined pipe sits in a position almost half the height of the digestor. The upper side of the both inlet and outlet chamber must be higher than the height of the top of the digestor in order to allow the slurry to flow into them when the gas pressure rises up.

3.1.1.4.3 *Digestor (Anaerobic digestor)*: This is the main part of the biogas plant where fermentation takes place and accumulates the biogas. This is essentially a circular shallow well dug under ground and coated with cement layer. The volume of the digestor may be designed according to the quality of biogass required.

3.1.1.4.4 *Movable cup and outlet pipe*: Movable cup is called safety cup which is fixed on top of the digestor. The gas outlet pipe is a short metal pipe fixed on the movable cup, a long flexible pipe is connected to it, so as to supply the gas for domestic needs such as cooking and lighting.

For speeding the fermentation in anaerobic digestor. The rotten organic materials should be first treated with compost. They mix in homogeneously with other matters and anaerobic bacteria. The important factor influencing the living condition of anaerobic bacteria is to keep a suitable temperature. The temperature range is from 10-65°C. In addition, pH should be maintained within a range of 7 to 8.5 once all the parameters are followed, raw material undergoes fermentation process and biogas is produced. It is essential to pay attention to alkalinity and acidity of the fermentation.

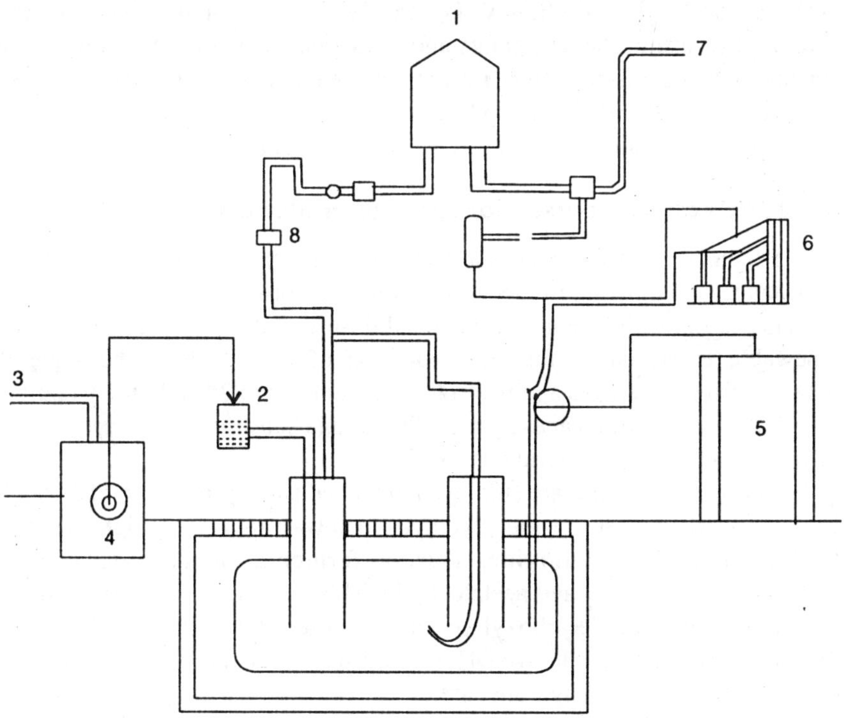

1. Gas tank
2. Conditioning tank
3. Feed tank
4. Feed pump
5. Sludge tank
6. Solar collector
7. Excess gas
8. Desulfurisation vessel

Fig. 3.2. Pilot plant for biogas production

When organic materials are fermented by anaerobic digestion, biogas is produced. The residue and dung solution are both excellent organic manure.

During the anaerobic reaction, the organic materials are gradually decomposed and converted into methane, carbon dioxide and other gases. The insoluble and decomposable organic and inorganic solids are mixed in the dung residue.

3.2 PRODUCTION OF ALCOHOL AS A BIOFUEL

According to Oxford dictionary 'alcohol' is a colourless inflammable liquid. It can also be an intoxicating agent in beverages and now being considered as a potential automobile fuel. In recent years, the use of alcohol as biofuel has been of considerable interest. The petroleum reserve is dwindling drastically and crude oil price is increased at alarming rate in international market. Hence, alcohol can be used along with petrol in automobiles as a potential biofuel. Although worldwide production of ethanol touches 1200 million tons during early 90's and US production was 70% and is being the largest producer of alcohol by synthetic means (Table 3.2 and 3.3). The growing demand for ethanol is mainly from industrial sector like pharmaceuticals, pesticides, paints and academic areas. In order to meet the demand, more emphasis is to be given to utilise biomaterials (Maize, potatoes, *cassaua* and *sorghum*) for the production of alcohol. However, several successful attempts have already been made in this direction.

3.2.1 Principle Biomass for Alcohol production

The principle plant biomass substrate for the production of alcohol are canesugar, starch, cellulose and hemicellulose. Substantial amount of sugar is locked up in cellulose and in hemicellulose as an immense storehouse of renewable energy stock for the production of fuel alcohol. The sugar is obtained from sugarcane and beetsugar, contains up to 20%. In the Brazilian approach sucrose from canesugar is a favourable substrate for the production of fuel alcohol by yeast fermentation.

Primarily yeast strains are employed in the industrial production of alcohol. They convert glucose by the glycolytic pathway to give a high yield of ethanol and carbon dioxide. During fermentation puruvic acid is decarboxylated to CO_2 and acetaldehyde. This conversion is carried out by an enzyme pyruvate decarboxylase. In the subsequent step, acetaldehyde is reduced to ethanol while NADH is oxidised to NAD^+. This reaction is catalysed by alcohol dehydrogenase.

$$C_6H_{12}O_6 \rightarrow 2C_2H_5OH + 2CO_2$$

Theoretically yield of alcohol from glucose is calculated to be 51% by weight. During fermentation pH is maintained below 5 to encourage yeast fermentation.

3.2.2 Raw Materials

Making ethanol by fermentation requires raw materials such as cereal grain (corn, wheat, barley, *sorghum*) sugar cane (molasses), sugar beet, fruit processed residue, starch crops such as potatoes, rice, dairy waste, paper industry waste, and materials as wood and plant cellulose, hemi cellulose etc.

3.2.2.1 *Ethanol from starch*: A variety of starch sources have been proposed for the production of fermentable ethanol, such as grains, *cassava*, sweet potatoes and sweet sorghum.

3.2.2.1.1 *Starch saccharification*: In order to release glucose from the starch, saccharification is done by enzymatic process and acid hydrolysis of starch. Two enzymes are used for enzymatic degradation of starch. One is a-amylase causes random hydrolysis, a-linkage of starch. This enzyme slowly converts starch into maltose, glucose and some oligomeric dextrins. In acid hydrolysis, breakdown of glucose is accompanied by further degradation of sugar to formic acid and livulic acid.

Table 3.2. Ethanol yield by various crops

Crops	Ethanol yield (L/tons)
Sugar crops	
Sugarcane	70
Sugar beet	95
Sweet Sorghum	86
Molasses	280
Cellulosic Material	
Wood	160
Starch crops	
Cassava	180
Sweet potatoes	125
Corn	370
Wheat	355
Potato	100

Table 3.3. Energy inputs into industrial phase of ethanol production

Feed stock	Country	Energy Inputs (MJ/Kg) Ethanol
Sugarcane	USA	16.0
Sugarcane	Brazil	18.5
Sweet Sorghum	Brazil	23.0
Cassave	Brazil	25.4
Cory	USA	19.4

3.2.2.2 *Ethanol from cellulosic material*: The main source of cellulosic materials are waste from agricultural field, municipal waste, industrial waste from silviculture. In order to facilitate free access for the degradation some pre treatment is done.

3.2.2.2.1 *Steam treatment*: Steam explosion is given to the green wood chips to 200°C for 10-30 minutes. During this treatment, acids are formed by the decomposition of hemicellulose under high temperature and pressure.

3.2.2.2.2 *Solvent treatment*: Appropriate solvent is used to remove cellulose from the matrix. The widely used solvent is cadoxen. It is an aqueous alkaline solution of ethylenediamine and cadmine oxide. After cellulose solubilisation, addition of excess water to the solution causes precipitation of the cellulose, which is then removed by hydrolysis.

3.2.2.2.3 *Acid hydrolysis*: Commercial operations involve acid hydrolysis of cellulose for the release of fermentable sugar from the native feed stock. Crystalline cellulose is soluble in 72% H_2SO_4 or 42% HCl at 10-45°C. The polymer is depolymerised to oligosaccharides. The oligomer mixture is then diluted to a lower concentration and heated to about 100-200°C for 1-3 hr. This step converts oligomeric starch chains to their monomeric glucose residue.

3.2.2.2.4 *Enzymatic*: Enzyme cellulose is capable of hydrolysing cellulose. B. glucosidase hydrolyse cellobiose and short chain oligosaccharides to glucose.

3.2.3 Gasohol Experiment

Gasohol is a mixture of alcohol and petroleum upto 20%. It is used as a fuel in automobiles, primarily as unhydrous alcohol. The ethanol blending has improved performance of engine. Brazil was the first country to use and employ gasohol experiment in automobile. It is the second largest producer of sugarcane in the world. The sugarcane waste generated is diverted towards alcohol production by fermentation. Petrolblend ethanol exhibits a higher economic value than ethanol as direct substitute. United States also used blended ethanol with petrol up to 5%. In U.S.A some amount of gasoline is blended with 10% ethanol. They produce corn as one of the main crop and is obviously the main biomaterial for alcohol production. The over production of corn reduces the prices and promotes to transform surplus corn into ethanol by fermentation. Several European countries including France are exploiting sugar beet, sugar cane, cassava for the production of alcohol. Use of gasohol reduces air pollution. Ethanol produced by corn fermentation reduces carbon monoxide level. A study shows that carbon dioxide levels could be reduced by 20% if vehicles especially car uses alcohol as fuel. The degree of investment in some countries for alcohol production increased to several folds. Sweden, Kenya, US have made quantum investment in alcohol production. In Brazil, government spends about 15 billion novo. cruz every year, for giving subsidy to alcohol production from sugarcane. This is followed by Zimbabwae, which launched a massive programme on alcohol production.

Alcohol production in rural areas can provide more jobs. It can be considered as income generation for people. The potential benefit of national alcohol programme has been stressed by Brazil, since its implementation in 1975. Farmers were encouraged to set up alcohol distillation unit. In Brazil, for several political and economic reasons a national ethanol production gained momentum to increase ethanol production from sugar cane. The Government has adapted number of approaches in its ethanol development plan to try and spread social and economic benefit of the rural sector.

3.3 MICROBIAL HYDROGEN PRODUCTION

Hydrogen is a clean burning fuel. It is the gas produced by numerous bacteria. The organism *clostridium perfringes* produces large quantity of hydrogen gas during fermentation of carbohydrate. Hydrogen producing organism possess an enzyme hydrogenase, which facilitate joining of two hydrogen ions and two electrons to form one molecule of hydrogen. In presence of light, enzyme hydrogenase separates energy rich electron from ferrodoxin and transfer to hydrogen ion and ultimately hydrogen gas is produced. In addition, Hydrogen gas can also be produced by degrading complex organic matter using mixed culture of bacteria. In an interesting approach photosynthetic bacteria are involved in hydrogen production.

The technique of hydrogen production using anoxygenic phototroph bacteria has been highlighted by various workers. Organic waste with low nitrogen contents are a suitable substrate for hydrogen production. The waste used as electron donors for photo production of hydrogen include sugar refinery waste, paper mill waste and, distillery waste water and also the clarified slurry of biogas plant.

Microorganism for hydrogen production: The production of hydrogen is performed by variety of microorganisms. These are *Klebsiella pneumoniae, Cellulomonus* sp., *Clostridium butyricum, Pseudomonas fluorescence, Chlamydomonas reinhardtii* and *Synchococcus cedrorum*. The enzyme hydrogenase has been obtained from 15 species of bacteria and algae.

Immobilised cells of an apparently contaminated cultures of *Rhodo-Pseudomonas rubrum* were found to produce hydrogen from glucose. The substrate glucose is derived from hydrolysed cellulose. The combination of photosynthetic bacteria with anaerobic bacteria could provide a suitable system for hydrogen photo production from carbohydrate.

The anaerobic bacteria produce organic acids which are then converted to hydrogen by photosynthetic bacteria.

In cyanobacteria nitrogenase enzyme helps in the production of hydrogen gas. This enzyme reduces nitrogen to ammonia with associated release of hydrogen gas. In an another system, combining algae *chlamydomonas reinhardtii* with *R.rubra* have been shown to produce hydrogen gas. The entire process was carried out under light and dark cycles. In the light green algae accumulated carbohydrate, which was fermented in the dark period to produce organic compounds such as acetate, ethanol, glycerol and formate. Among these formate was utilised for hydrogen production in dark by photosynthetic bacteria.

3.3.1 Bioreactor

Several bioreactors are available for photohydrogen production. All the reactors designed so far have the source of illumination outside the reaction vessel. It is made up of acrylate or polycarbonate. Typically one design glass reactor consists of chamber between two concentric cylinder walls with an illumination provided by a bulb at the centre. In an innovative approach optic fibres are used to provide light to the reactor, Kobyashi and Yc (1986) developed a technique for outdoor production of hydrogen on a large scale. Outdoor hydrogen production was demonstrated in biosolar reactors. 150 litres of gaseous Hydrogen was produced per day/m^3 using anoxygenic phototrophic bacteria. Following are some of the simple bioreactors for hydrogen production in presence of light (Fig. 3.3 and 3.4).

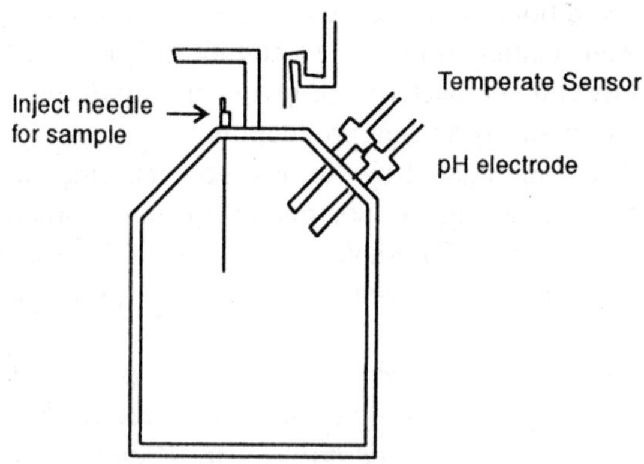

Fig. 3.3. Biosolar reactor for Hydrogen production

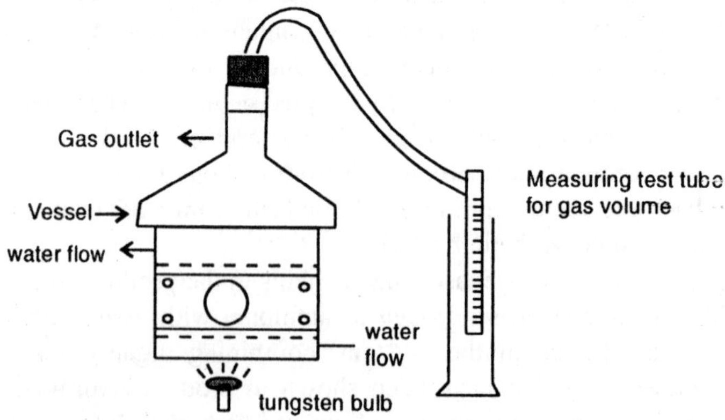

Fig. 3.4. Reactor for Hydrogen production by planar agar matrix

3.4 SOLAR ENERGY CONVERTERS

Chlorophylls are the green pigments in leaves is a superb solar energy collector. It captures incoming solar energy, converts it and then stores it in the plant tissue, particularly in leaves, wood etc. This energy can be recovered in a number of ways. The photosynthetic activity of living cells provides an opportunity to store solar energy for example, slowly growing tree produces firewood. Aquatic plants and microorganisms can store energy more rapidly. The pond algae grows in processed sewage, can be produced which is equivalent to 500 £ of algae per acre per day. This mass of organic material produced is a possible source of fuel. These are used to burn directly or by conversion to another form.

From algae to higher plants, contains a group of pigments known as "chlorophyll" pigments or photosynthetic pigments. The types of photosynthetic pigments are chlorophyll a, chlorophyll b. Xanthophyll and carotenoids. These pigments have unique capacity to trap solar radiation in the visible range between 400 to 700 nm. When sunlight falls on plant system, chlorophyll molecules are excited which results in transfer of energy from one chlorophyll molecule to another chlorophyll by resonance migration. Finally all the energy from chlorophyll is dumped into reaction centre molecule from where electron is ejected and enters electron transport chain. The energy rich ATP and NADP are formed during electron transport chain and are utilised in reducing atmospheric carbon dioxide into carbohydrate via calvin cycle. The energy based carbohydrate is stored in various parts in plant, including wood, root, stem, etc. The whole biomass is a feasible energy reserve. These are in turn utilised as a whole or as a waste to produce alcohol, biogas, fertiliser, etc (Fig. 3.5).

3.5 DEGRADATION OF CELLULOSE

Cellulose is one of the major component of plant biomass. It is one of the important renewable source of raw material. Cellulose in association with lignin forms lignocellulose. It makes up about half of all matter produced through photosynthesis and represents over 90% of the dry weight of the cell. Lignocellulose consist of 3 polymers. They are cellulose, hemicellulose and lignin binding with each other. On an average lignocellulose consists of 45% cellulose, 30% hemicellulose and 25% lignin. The global reserve of cellulose comes from non woody aquatic plants and natural fibres like cotton. Their reserve runs into million metric tons in energy plantation crops, such as sugar cane, sugarbeet and sorghum. The high cellulose content plants like kernel and elephant grass and high starch containing cassava plants are exploited in the production of industrials based compounds. The sugar cane bagasse contains high cellulose. If the agriculture production exceeds the required food demand, the surplus cellulose waste residue can be exploited for the production of fossil fuel.

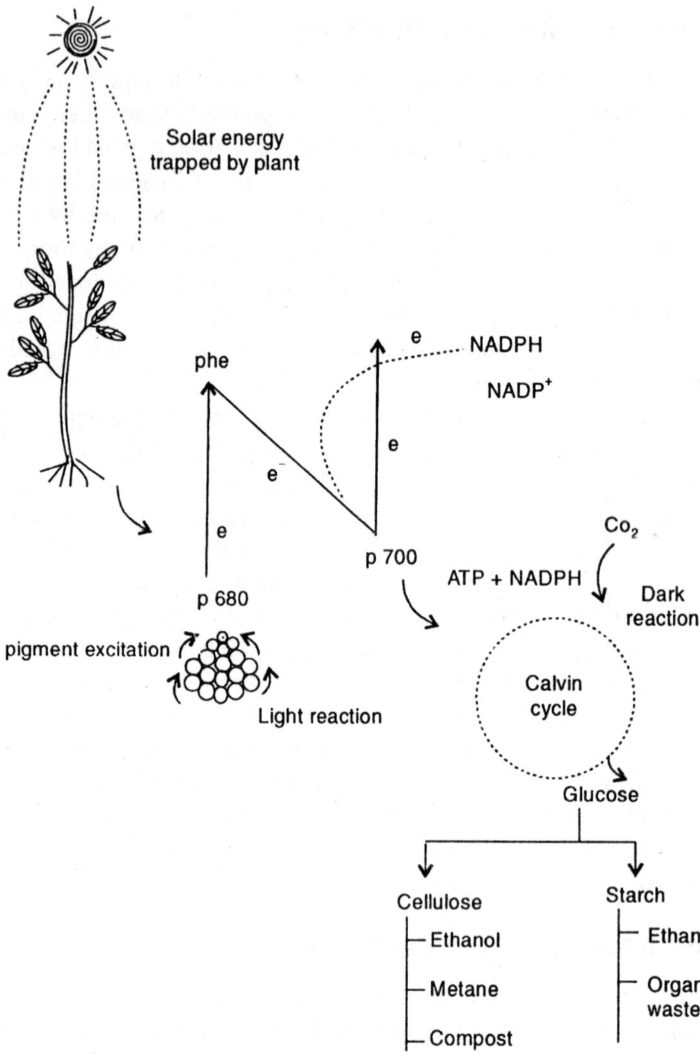

Fig. 3.5. Conversion of solar energy into useful products

The energy rich cellulose is locked in biomass. It can be subjected for microbial degradation to release more simple sugar like glucose. The enzymatic degradation of cellulose is a dynamic process. The enzyme known as cellulose can digest cell wall material and is able to break down cellulose polymer into small sugar molecules, which are permeable to bacterial cell. The enzyme cellulose can remove lignin barrier.

3.5.1 Microbes for cellulose degradation

Hundred of species of fungi and bacteria are capable to degrade cellulose. Among them, extensively studied, cellulotic enzyme sources are *Trichoderma and Cellulomonas and Clostridium thermocellum.*

Trichoderma reesei produces three types of cellulose depending enzyme like endogluconase, cellobiohydrolase and β-glucosidase. Four genes coding for two endoglucanase (EGI and EGII) and two cellulobiohydrolases i.e., CBH I and CBH II have been isolated from *Trichoderma reesei.* The endogluconases hydrolyse intermediate bond, along cellulose fibres. The ends are then attacked by another enzyme cellobiohydrolases with the release of disaccharide cellobiose.

In the hydrolysis, cellobiohydrolases enzyme begin to disrupt chain to chain interaction in the crystalline region of cellulose fibre. Finally, cellobiose is hydrolysed to glucose by β-glucosidase.

In recent years it has become possible to exploit cellulose for low cost recovery of hexoses and pentoses. Some successful attempts have been made to obtain cellulose derived sugar syrups which have wider industrial applications.

Process of Cellulose Degradation

Cellulase

↓

Endogluconases
+
Cellobiohydrolases

Cellodextrins + cellobiose

$$\text{Cellodextrins} \xrightarrow{\beta\text{-glucosidase}} \text{Glucose}$$

$$\text{Cellodextrins} \xrightarrow[\text{or cellobiohydrolases}]{\text{Endoglucanase}} \text{Cellobiose}$$

$$\text{Cellobiose} \xrightarrow{\beta\text{-glucosidase}} \text{Glucose}$$

3.5.1.2 *Degradation of lignocellulose*: Lignocellulose consist of cellulose, hemicellulose and lignin, the first step in utilisation of biomass for fuel is, the release of carbohydrate and aromatic building block locked up in the structure of lignocellulose. Many fungi and bacteria equipped with enzymes are capable of degrading the polymers of lignocellulose. Many bacteria produce cellulose as well as enzymes capable of cleaving various bonds in hemicellulose. Once these polysaccharides are exposed by fungal delignification, further degradation is accomplished by fungi and bacteria.. Most useful enzymes for lignin cellulose degradation are lignin peroxidase, glucanases and xylonases. The genes for lignin peroxidase, cellulase and various hemicellulose degrading enzymes have been cloned and the enzyme is expressed in genetically engineered fungi and *Escherichia coli.* These advances hold promise for efficient utilisation of lignocellulose as a potential source of sugar substitute for fermentation process such as the production of ethanol.

3.5.1.3 *Degradation of Hemicellulose*: Hemicellulose are vast renewable source of sugar can be converted to ethanol. Hemicellulose is a polymer which constitutes 30% of total lignocellulose material. Xylans are the major hemicellulose in wood, where they account for 15-30% of the total dry weight. Fermentation method is required to convert xylan into alcohol. The structure of xylan is complex and their complete biodegradation requires the action of several enzymes. The enzyme xylanases isolated from many fungi and bacteria can degrade xylan.

The complete conversion of a glucoronaxylan requires endo-β-1-4 xylanase and β-xylonidase as well as β-glucoronidase and acetylxylanesterase.

3.5.1.4 *Commercial Significance of Cellulose*:

(i) Cellulose present in the biomass waste can be potential source for ethanol production.

(ii) The cellulose present in sugar bagasse is a good substrate for alcohol production.

(iii) It is a good substrate material for the microbial hydrogen production.

(iv) The industrial product sugar syrup is derived from cellulose which have wider industrial applications.

(v) Lignocellulose and hemicellulose present in the plant waste is a good source for biogas and ethanol production.

3.6 PETROLEUM PLANTS

Certain plant species accumulate high molecular weight hydrocarbons as a photosynthetic product. In the recent study, projections of world petroleum consumption and estimation of crude oil reserve indicate that the petroleum supply will be exhausted some time in the twenty first century. In order to combat future energy crisis. More emphasis is given to look for alternatives energy resources. Over exploitation of one particular non-renewable resources can be detrimental to mankind. Several attempts have been made to develop technology for the extraction of hydrocarbon (crude petrol) from plant sources, commonly known as petroplants. Petroreserve of plant is not a recent story. Malvin Calvin, a champion of calvin cycle, for the first time has thrown light on photosynthetically produced hydrocarbons from plants. He suggested that it could be a potential substitute for conventional source. Many decades back the use of *Euphorbia* as a source for gasoline was well known.

A comprehensive effort has been made to identify, classify and develope technology for petroplants presence of high molecular weight hydrocarbons can be converted into petroleum hydrocarbon. The petroplants can be grouped into essential oil, seed oil, laticiferous plants and resinous plants. In India, some of the research institutions like NBRI (National Botanical Research Institute) and Indian Institute of Petroleum, Dehradun launched a project on screening and cultivation of petroleum based plants. The studies had listed as many as 400 species belonging to six laticiferous families like Urticacae,

Convolvulaceae, Sapotaceae, Asclepidaceae and Euphorbiaceae. The future strategy involves cultivation of hydrocarbon rich plants, establishments of latex containing germ plasm bank, evolving technology for maximum biomass production and standardisation of hydrogenation of hydrocarbon type of material.

Some of the European countries have set up Gasoline refinery based on the supply of petroleum plants. For example, in Italy, Euphorbia gasoline refinery was established to exploit gasoline from vegetative hydrocarbon source. Some of the Euphorbian species yielded 5-20 barrels of oil per acre in a year. The latex obtained from *Calotropis procera* contains high hydrocarbon in ratio of carbon, hydrogen and oxygen in the hexane extract was found to be highly significant. The ratio of carbon and hydrogen in similar to crude petroleum oil. The algae biomass contains 30 percent of hydrocarbons closely resemble crude oil. Indian climatic conditions favours growth of a variety of petroleum hydrocarbon containing plants. Many vast barren land sources can be utilised to cultivate hydrocarbon rich petroleum plants (Table 3.4).

Table 3.4. Classification of petroplants

Seed oil plants	Laticiferous plants	Resinous plants	Algae plant
Ricinus communis	*Euphorbia antigonum*	*Pinus*	*Botrococcus*
Jatropha curcus	*E. nesifolia*	*Canarium*	*Chlorella pyrenoidosa*
pithosporum resiniferus	*E. nivulia*	*Vateria*	
calotropis sp.	*Asclepias*	*Dipterocarpus*	
sapium sp	*Calotropis gigantia*		
	Crytostegia grandiflora		

QUESTIONS

1. Mention various types of non conventional fuels. Add a note on their advantages over conventional fuels.
2. Discuss the significance of non conventional fuels in facing future energy crisis.
3. "Waste is a renewable source, Energy". Discuss.

   (BU Nov. 1999)
   (BU April 2002)
4. Give an account of Biogas production.

   (BU Nov. 1999)
   (BU April 2001)
5. Explain the role of methanogenic bacteria in Biogas production.

   (BU May 1996)

6. Explain how alcohol can be considered as renewable source of energy. Add a note on gasohol experiment.
7. Give an account of microbial degradation of Cellulose (Nov. 1999) (BU Nov. 1997)

SHORT QUESTIONS

1. Gasohol experiment (BU April 2001) (BU April 2002) (BU Nov. 1999)
2. Biogas (BU Nov. 1997)
3. Methanogenic bacteria
4. Petroplants
5. Solar energy converters
6. Microbial hydrogen production

4

Microbiology of Food and Water

4.1 FOOD

Food is a mixture of complex array of chemicals indispensable for maintenance of life. Generally food contains carbohydrate, proteins, lipids and vitamins. The same food that we eat may be responsible for illness. In terms of quality assessment, food need to be nutritious and metabolisable. Food is also a good substrate for growth and metabolism of microorganisms. They use food substrate to derive energy and also many essential molecules, which they cannot synthesize for its growth and development. There is an unprecedented relationship existing between science of food, microbes and biotechnology.

4.1.1 Food Spoilage

The spoilage of food is generally associated with rancidity, discolouration, drying out etc. Most of the food spoilage is a result of microbial activity, however, several other changes may not always be microbiological in origin. Storage of fresh meat, fish in open air encourages other the growth of microorganisms. Even aerobic storage of chilled meat either covered by thin film or uncovered may be suitable for the growth of psychrotrophic aerobes. Some of the bacteria found in flesh which are responsible for spoilage are *Pseudomonas, Acinobacter* and *Psychrobacter.* Gram-negative fermentation bacteria grows rapidly on meat. The presence of odours of volatile esters, alcohols, ketones confirms the predominant role of *pseudomonas* in food spoilage. Several sea foods like lobsters, crabs, molluscs carry good number of bacteria in their skin, gills and alimentary tract. They are mainly gram negative *Pseudomonas, Vibrio, Flavobacterium* etc.

4.1.1.1 *Spoilage of fresh fish:* The spoilage of chilled fish is mainly due to gram negative *Shewanella putrefaciens* and *Pseudomonas* sp. etc. The bad odour is due to decomposition of fish, which generates hydrogen sulphide, methylmercapton and dimethylsulphide.

4.1.1.2 *Spoilage of plant product:* Spoilage of fruits and vegetables is generally caused by fungi and bacteria. Fruit juice with their low pH gives considerable protection against bacteria but are susceptible to fungi. However, many vegetables have high pH and can be susceptible to bacterial spoilage. The most common mould on fruits and vegetables is the grey mould *Botrytis cinera.* The storage of harvested food grains especially cereals are predominantly spoiled by mould. They are well adapted to low water condition. The important fungi are *Penicillium* and Aspergillus. The mould attack reduces germination potential of cereals and other seeds. During infection mycotoxins are produced followed by discolouration and musty odour production. All these processes are ultimately terminate in complete decay of food grains. Several lipolytic moulds like *Aspergillus niger, Penicillium* and *Rhizopus* can grow on oil seeds and produce mycotoxins. The oil extracted from the seeds contain free fatty acids, which show higher rancidity due to oxidation.

4.1.2 Food poisoning

Common sources of food poisoning include beef, chicken, turky backed goods. *Staphylococcus aureus* enter food through various routes and causes food poison. About 50 percent of this bacterial strain produces enterotoxin. It causes nausea, vomitting, diarrhoea.

4.1.2.1 *Salmonella food poisoning:* Salmonella are the most frequent causes of food borne disease in several parts of the world, especially in USA. The common food sources for this organism are fresh pork and chicken. Slaughter house encourages the spread of *Salmonella* to proliferate in the milk. Symptoms are nausea, vomitting and diarrhoea.

4.1.2.2 *Botulism food poisoning:* Botulism is a severe intoxication caused by ingestion of botulism toxins. Usually backery foods, fish, fruits, vegetables and canned foods are the chief source for *Clostridium botulinum.* The toxin produced by this bacteria is the most poisonous and kills the host by damaging nervous system. The ingested neurotoxin cannot be destroyed by enzymes of digestive tract and it is easily absorbed into the circulatory system. Symptom appears within 12 to 42 hrs after the consumption of the contaminated food. The main symptoms are double vision, muscle paralysis, vomitting etc.

4.1.2.3 *Food poison by virus*

4.1.2.3.1 *Rotavirus:* Rotavirus responsible for infant diarrhoea is disease. The viruses are carried by mice, rabbits, pigs. The symptoms of rotoviral infection vary from mild diarrhoea to severe gastroenteritis. Death can result from severe dehydration and electrolyte imbalance.

4.1.2.3.2 *Norwalk virus:* Norwalk virus is the prime agent in causing viral gastroenteritis, and epidemic diarrhoea. The symptoms includes vomitting, diarrhoea and low fever. The illness is predominant during colder months. These viruses are generally transmitted through fecal-oral route. However, patient recovers after few days without any treatment.

4.1.2.3.3 *Viral Hepatitis:* Hepatitis is infectious. It can be seen in humans, monkey. They are mainly transmitted by fecal-oral route and by contaminated food.

4.1.3 Preservation of Food

In order to ensure quality and complete safety for consumption, several preservation methods have been in practice. During food preservation several factors which are responsible for microbial growth are taken into consideration as a basis of preservation. Following are some of the methods used for food preservation (Table 4.1).

4.1.2.1 *Radiation treatment*: Irradiation is one of the most effective way of preserving foods. There are three types of radiation treatment, considered the microwave radiation, UV radiation and gamma radiation.

4.1.2.2 *Microwave*: In the electromagnetic spectrum the microwave occupies 109 Hz frequency for food preservation. The low quantum energy 2500 MHz is employed. In domestic houses, microwave can be used to preserve our foods. In principle, Microwave generates or imposes temperature on water molecule present in the food leads to vigorous oscillation. Kinetic transfer to surrounding molecule leading to rapid rise in temperature. Microwave has been used to defrost frozen foods. In addition, it can also be used to pasteurise high acid foods.

4.1.2.3 *UV Radiation*: Ultraviolet has wider application in food preservations. Ultraviolet can damage DNA of microbes contaminated food. Resistance to UV light by microorganisms are dependent on the ability to repair the damaged DNA. Bacterial spores and mould spores exhibit higher resistance to UV radiation. Inside meat storage room low pressure mercury vapour discharge lamps are used and their emission is at a wave length of 254 nm which is adequate to monitor microbes.

4.1.2.4 *Ionising Radiation*: High energy ionising radiation can ensure high degree of protection to the food. Ionising radiation include X-rays and Gamma rays. They can affect microorganisms directly by interacting with molecules within microbial cells. They produce hydroxyl radicles which can cause severe damage by single and double strand break in DNA molecule. The dosage of radiation is important as it involves health risk. The joint expert committee under World Health Organisation and International Atomic Energy Authority had set up standard parameters for food irradiation. They recommended food irradiation upto 10 KGy and held the view that irradiated foods does not require any further toxicological test. Some pathogenic clostridium botulism spore exhibit resistance to radiation. So high dosage is required to achieve good result with 12 D (45 KGy) low acid foods. Radiation treatment had successfully eliminated *Salmonella* from meat and chicken and this increased the shelf life.

4.1.2.5 *Heat treatment*: *Pasteurization*: Pasteurization is a heating process in the range of 60-80°C and applied upto a few minutes. Pasteurization can ensure safety of milk, ice cream mix, etc. The process can eliminate large properties of spoilage organisms. Certain spore forming organisms like *Microbacterium, Enterococcus* can survive pasteurisation process.

4.1.2.5.1 *Appertization*: Appertization is a sterilisation process for canned foods. The appertised products have a longer shelf life. Any organism that survives appertization process are non pathogenic and incapable of growth under normal conditions of storage. The canned food in temperate countries are safe due to low temperature and spores of *Thermophiles* will not multiply under extreme cold condition.

4.1.2.6 *Chemical preservation*: Addition of certain chemicals to food during preservation can prevent to spoilage. Preservatives are chemicals capable of arresting microbial growth in food. Benzoic acid is proved to be effective antimicrobial preservative. It inhibits the growth of moulds and yeast. Parabens are hydroxybenzoic acid esters used to preserve bakery and fruit products. Acetic acid is more effective preservative for pickles and sauces. Sorbic acid is active against yeast and moulds and also against *Clostridium* spore germination. Nitrite is a potential food preservative against wide range of microorganisms like *Salmonella, Flavobacterium, Lactobacillus and Clostridium botulinum*. The sulphur dioxide, a highly reactive molecule can inhibit microbial metabolism, damage disulphide linkage in proteins.

4.1.2.7 *Freeze preservation*: Preserving food at subzero temperature can provide maximum safety to food. The ideal low temperature for preservation is between –1 to –3°C. At freezing temperature water begins to transform into ice in inter as well as intracellular region of the cells. Bacterial spores can survive low freezing temperature with gram positive strains showing greatest resistance.

Table 4.1. Different preservation methods

Irradiation	Heat process	Chemicals	Low temp.
(a) Microwave radiation (b) Ionising radiation: Forms hydroxy radicals and damage DNA (c) UV radiation: Damage DNA of microbes	(a) Pasteurisation: 60°-80° for few minutes (b) Appertization: For canned foods	(a) Benzoic acids: Effective against moulds, yeasts (b) Acetic acid (c) Parabens (d) Sorbic acid (e) Nitrite	–1 to –3°C or –1 to –18°C

4.1.4 Microbiological Examination of Food (Microbial Quality of food)

Microbiological examination of food is indispensable to determine the presence of pathogens. Following are some of the methods for microbiological examination of food or microbiological quality of food.

4.1.3.1 *Direct Examination:* A small sample is directly observed under dark field or bright field microscope. The organisms can be heat fixed and stained for observation. The direct epifluorescent filter technique (DEFT) has been developed for rapid determination of microbiological quality of farm milk. The concentrated bacteria is stained with acridine orange and counted directly under epifluoresence microscope.

4.1.3.2 *Culturing organisms*: Several selective and elective media were designed to encourage rapid growth of specific microorganisms or a group of microorganisms. Some media can inhibit the growth of other organisms when designed specifically. *Clostridium perfringens* can be grown on cooked meat broth at 44°C. The Baird-Parker agar is used for presumptive isolation of *Staphylococcus aureus*. This media contains lithium chloride and sodium tellurate as selective chemicals and elective agents are glycine and sodium pyruvate. The list of media specific for microorganisms are given in Table 4.2.

Table 4.2. Types of media for specific organism

	Media	Microorganisms
(i)	Broth MacConkey	Coliforms in water
(ii)	Baird-Parker agar	*Staphylococcus aureus*
(iii)	Thiosulphate, Bile, Citrate agar	Vibrios
(iv)	Brilliant green lactose bile broth	Coliforms in food
(v)	Crystal violet, Azide blood agar	Faecal Streptococci

4.1.3.3 *Rapid test for specific organisms and Toxins:* Several novel techniques are available to determine specific organisms and toxins. Some of the rapid tests are ELISA and PCR. In immunological method ELISA is prominently used for wide variety of samples.

4.1.3.3.1 *Immunological method:* Detection of organisms by immuno-technique involves, antibodies raised against specific antigens of micro-organisms. In order to detect toxins, antibodies have been raised to specific toxins of botulinum. Fungal toxins can be detected by slightly altered method. Mycotoxins are haptens and are non immunogenic. It can be made immunogenic by binding some chemicals to carrier proteins and antibodies are raised against aflotoxins. The essential features in immunoassay is, binding of antibodies to antigenic proteins.

4.1.3.3.2 *Nucleic acid probe*: Nucleic acid probes are designed and prepared to determine and hybridise to specific regions of chromosomal or plasmid DNA or to RNA. To facilitate hybridisation of probe DNA with sample DNA, sample is electrophoresed followed by southern hybridisation. Finally autoradiography is done to detect the sequence hybridisation.

4.1.3.3.3 *PCR*: Polymerase chain reaction is fastest and accurate method of organism detection. In the PCR method specific fragment of organism DNA is amplified and electrophoresed to detect the presence of organism. Availability of primer can decide successful application of the method.

4.2 MICROBIOLOGICAL QUALITY OF WATER

Water is generally contaminated by bacteria. Release of faecal contaminated sewage water into the streams and rivers is one of the main routes of contamination and it is of major concern. The major water contaminating microorganisms include pathogenic strains of *Escherichia coli* and *Streptococcus faecalis*. These two are main bioindicators of potability of drinking water. *E. coli* is gram negative, non-spore forming rod shaped bacterium and belongs to enteric forms which can pollute water through faecal contamination. The *Streptococcus faecalis* is gram positive enterococcus forms. To determine the safety of drinking water microbiological quality of water is to be done. The presence of coliforms can be determined by presumptive test, conformation test and completed test.

4.2.1 Presumptive Test for coliforms

Presumptive test is done for water sample to confirm lactose-fermenting gas producing bacteria. It is used to determine the most probable number (MPN) of coliforms in a sample of water. The method is based on inoculating replicated tubes of liquid medium with three different sample size and number of positive tubes and each sample is determined after incubating the tubes. The requirement for MPN are (i) Lactose broth of single and double strength (ii) Durham's tubes (iii) Incubator and others. Arrange three sets (3 each) tubes containing double strength lactose broth. Each set contains 10 ml, 1.0 ml and 0.1 ml of lactose broth. Dispense 0.1 ml, 1 ml and 10 ml of water sample to set 1, 2 and 3. Incubate all tubes at 37°C for 24 hrs. for gas production. The production of gas confirms coliforms in the sample.

4.2.2 Confirmed Test

Confirmed test can differentiate coliforms from non-coliforms as well as gram positive and gram negative bacteria. In the test, EMB agar plates are inoculated by sample taken from previous positive tubes. The EMB agar media, which contains methylene blue that inhibits coliforms. Based on the size of dark centered colonies and metallic sheen, *E. coli* and *E. aerogenes* can be differentiated. *E. coli* colonies on EMB agar medium are small and appears as metallic sheen where as *E. aerogenes* colonies are large and non metallic.

4.2.3 Completed Test

In the completed test colonies grown on EMB agar plates is removed and inoculated into lactose broth. After incubation for 24 hrs at 37°C, if broth exhibits gas production and acid, presence of coliforms is confirmed (Fig. 4.1).

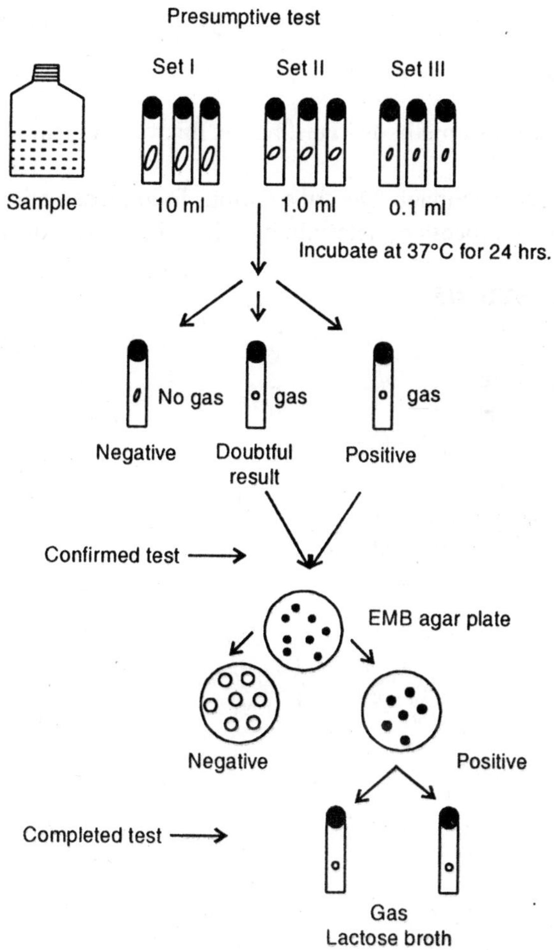

Fig. 4.1. Tests for coliforms in water.

4.2.3.1 *Detection by membrane filter:* This is one of the quick method for coliform test. Water sample is passed through a membrane filter of 0.45 micrometer, with a porosity of 150 micrometer in thickness. Bacteria is retained on the filter. Place a membrane on pad saturated with Endo MF medium and incubate at 37°C for 24 hrs. Coliforms develop with golden metallic sheen.

4.2.3.2 *Disinfection of drinking water:* In order to maintain the quality of drinking water fit for human consumption, water is drawn from various sources (rivers, lakes, streams) and collected in wide oxidation ponds. In the initial pretreatment water is allowed to pass or filter through sand mixed with gravel or pebbles. This would trap or retain macro particles. Particle free water is then subjected for aeration followed by spraying bleaching powder of approximate concentration. Chlorination of water would kill micro-organisms. and water is recycled for better aeration. This would increase oxygen content of water. Later purified water is stored in tanks.

QUESTIONS

1. Discuss various methods in testing the microbiological foods. (BU April 2000)
2. What is food spoilage. Describe various food preservation methods.
3. Explain the process of determining microbial quality of water.

SHORT QUESTIONS

1. MPN (BU April 2001)
2. Food spoilage
3. Food poisoning
4. Food preservation by irradiation
5. Test for coliforms
6. Heat treatment
7. Food preservation by chemicals

5

Treatment of Municipal Waste and Industrial Effluents

All metropolitan cities and major cities are generating millions tonnes of garbage every day. Rapid urbanisation, dense population have contributed to a greater extent in garbage production. Generally garbage is a municipal waste. Some places like restaurants, choultries, temples, hostels, hospitals are to main waste generating centres. These wastes are the breeding centres for mosquitoes, flies and even harbors several pathogenic microorganisms. Domestically produced waste food materials may enter drainage and flow in the form of sewage. Municipal waste can be grouped into two types — (1) solid waste and (2) liquid waste. Solid and liquid wastes consists of bio-degradable and non biodegradable wastes. Major composition of municipal waste is organic matter (10%) plastic (25%) and other matter (approximately 50%).

5.1 MUNICIPAL WASTE TREATMENT

Several simple biotechnological methods have been suggested and implemented for waste management. Following are some of the methods of municipal waste treatments.

5.1.1 Waste Collection

Municipal garbage is to be collected from waste generating centres. Door to door collection of domestic garbage is an ideal choice. In 2000, Bangalore City Corporation had successfully implemented a novel door to door garbage collection programme — "Swaccha Bangalore" meaning Clean Bangalore.

5.1.2 Waste Disposal

Disposal of accumulated waste is done regularly. Municipal waste, once collected are transported to countryside and disposed in waste land. Once non-biodegradable materials like plastics, tin, rubber are separated, the remaining waste is burnt to ash.

5.1.3 Compost from Municipal Waste

Composting of organic waste is an age old practice in rural areas. Collecting waste materials like leaves, slender dried stems, weeds and mixing them with dung in a pit to prepare compost for manure is well known to villagers Development of improvised methods resulted in better waste management like composting and vermicomposting. Compost is an organic manure prepared by the action of microbes on waste by biochemical process. During composting process the material undergoes intensive decomposition in high temperature and Moisture in a pit. The high temperature which is generated in the later period during decomposition would destroy harmful pathogenic bacteria and fungi in garbage. It takes 4-6 months to obtain black to brownish powdered material known as compost. Decomposition is associated with aerobic (which requires oxygen) and anaerobic (absence of oxygen) process. The compost prepared by improved techniques are more stable and improvises soil fertility to a greater extent. However, several parameters are indispensable to maintain high quality compost. They are high C:N ratio, particle temperature and blending.

5.1.3.1 *Materials for compost:* In view of better urban waste management, following are the essential prerequisites for compost preparation.

5.1.3.1.1. *Solid Municipal Waste:* These are mainly collected from towns and cities which are the main waste generating centres. Solid waste are waste vegetables, food materials from houses, canteen and restaurants, etc.

5.1.3.2 *Solid garden waste:* Garden waste includes dried leaves, grass, weeds, stubbles etc.

5.1.3.3 *Liquid waste:* Liquid wastes like sewage water, sludge collected from sewage water can be employed for compost preparation on a larger scale.

5.1.4 Method of compost preparation

General method of compost preparation is described as follows:

(a) Compost preparation requires a deep pit or heap which is 6-8 feet in breadth, 10 feet in length and 3 feet in depth.

(b) Waste materials brought from different places is spread evenly into the pit in layers of 3 inch thickness.

(c) Above this layer cattle dung and soil is spread (3 inch) evenly. Water is sprinkled to maintain moisture. Addition of dung acts as a catalytic agent.

(d) Within few days 3/4 of the pit should be filled with the material layer by layer and one fourth of the pit is left empty to facilitate turning.

(e) Water is sprinkled twice a day to maintain 100% moisture.
(f) The whole material (heaps) in the pit is turned once in a week. It is immediately followed by sprinkling water at the time of first, second and third turning. The second turning is done after 15 days of the first turning and the third turning is usually done after 2 months. The whole system is left for decomposition for a month. The turning of heaps facilitates not only circulation of air and water but also facilitates of decomposition by bacterial activity (Fig. 5.1).
(g) After 20 days to 1 month the materials get contracted and becomes more compact and sinks down the tank by 20-30 inches. Hence filling is repeated again and the pit is sealed using mud paste.

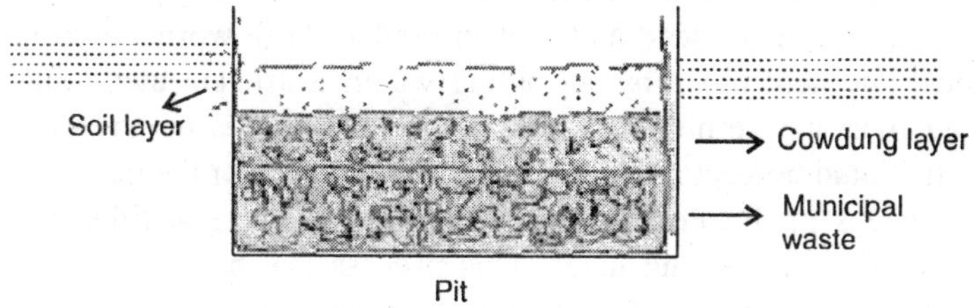

Fig. 5.1. Pit for compost.

5.1.4.1 *Microorganisms for compost:* Several cellulose and lignin degrading bacteria enhance the composting process in terms of quality. The cultures. *Trichoderma viridae, Aspergillus awamari, Trichums spiralis* were added during preparations. The presence of phosphate solubilising nitrogen fixing bacteria increases the nutrient quality of compost.

5.1.5 Vermicompost

Vermicompost is an organic manure prepared by using earthworms. Soil fertility is increased by the activity of earthworms. A good number of earthworms in soil is an indication of soil fertility. The earthworm is a friend of farmer. Earthworm is an excellent manure for vegetation and garden plants. It can be prepared easily. Earthworms can be used in a systematic way to convert household kitchen waste to compost.

5.1.5.1 *Earthworm in Compost:* Charles Darwin first demonstrated the ability of earthworm to breakdown vegetable matter and incorporated it into the soil. Since then considerable interest was shown in the field of vermitechnology. Earthworms are important agents in breaking down plant remains, vegetable waste and garbage into small pieces, which provide greater surface area for further microbial activity. It improves soil aeration, porosity and water holding capacity. The carbon to nitrogen ratio (C:N) has been reported to be higher in worm cast, than in the soil. Earthworm can increase the growth of plant in soil. The most popular species of earthworm for compost preparation is *Eisenia foetida* which is involved in breakdown human, animal, vegetables, industrial organic waste and transform them into high quality compost. The Indian species, *Perionyx excavata, Octochaetona serrata* are the most common.

The beneficial role of earthworm in treating municipal sludge have been well documented. The earthworm can digest and exhibits high tolerance towards heavy metal concentration in the sludge.

5.1.5.2 *Materials for vermicompost:* Organic materials for vermicompost, primarily consists of household garbage and urban garbage such as fruit, vegetable trimmings, flowers, coffee and tea waste, tea leaves, grains, hair, spoilage food etc.

5.1.5.2.1 *Methodology:* The vermicompost pit or tank is made up of 2.5 m × 1.5 m × 1 m. At the bottom vermibed is prepared by placing loamy soil above a thin layer of sand and broken bricks. Earth worms are released into this bed and kept moist by sprinkling water. Sufficient cattle dung are then placed over the vermibed. Above which dry leaves or hay is spread unevenly. It should be kept moist and watered regularly for the next 4 weeks. After four weeks of decomposition urban garbage waste is filled up over partially decomposed organic layer. This process can be repeated twice a week. The pit has to be moistened by regular watering. The top of the pit is covered by mesh to protect the earthworms from birds. Aluminium or asbestos sheet should be placed above the earthworm pit to provide proper shade.

The burrowing activity of the earthworm, fast movement between different layers, make horizontal burrowing and deposit the cast increases mixing process. Simultaneously earthworms breakdown garbage waste into smaller particles and facilitate further microbial decomposition. After 45 days the whole mass is transferred into black amorphous, humus soil known as vermi compost (Fig. 5.2).

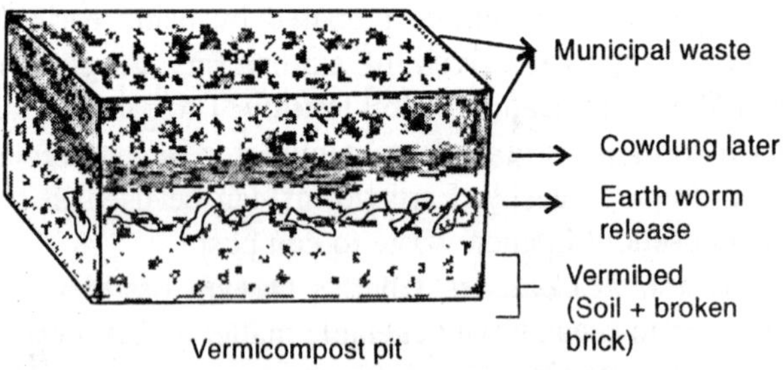

Fig. 5.2. Pit for vermicompost preparation.

5.2 TREATMENT OF INDUSTRIAL EFFLUENTS

Water is a unique solvent and a good carrier. It gets contaminated easily. One of the most common source of water pollution is the discharge of untreated sewage into water. Industrial effluents and agriculture discharge into water may result in severe water pollution. The industrial effluents are mainly produced by chemical plants, mills, distilleries, tanners, detergent and plastic industries. Discharge from domestic waste, in the form sewage enters water stream. Detergents pollute water to a greater extent as phosphates are the major components of detergent. They would favour luxurious growth of algal bloom which consumes more oxygen, as a result of which oxygen content of water gets reduced. Once oxygen is depleted in water its biological oxygen demand (BOD) increases, which is detrimental to living organisms.

Organic and inorganic pollutants are present in effluents released from breweries, dying textiles, tanneries, paper pulp and other industries. Sulphuric acid (H_2SO_4) from coal mines is a serious pollutant, that corrodes concrete material. Industries which includes chemical fertiliser units discharge pollutant like sodium, copper, chromium, cadmium, mercury, lead, zinc as metallic effluents (Table 5.1). Modern agriculture methods involve extensive use of pesticides and herbicides in order to increase crop yields. Agriculturists depend upon chemical fertilisers, fungicides, pesticides and herbicides. Ultimately these chemical pollutants enter water source as agriculture pollutants. Pesticides are hazardous toxic chemicals which enter food chain and create serious environmental problem. Pesticides like BHC (benzene hexachloride), DDT (dichloro diphenyl trichloroethane) and PCB (poly chlorinated biphenols) and others, due to indiscriminate usage, enter water source and pollute it. The pollutants from water may enter into plants, fishes, birds and even man.

5.2.1 Effluent treatment

Treatment of industrial effluents is more complex than treatment of domestic waste. Biological treatment could be well suited for waste from dairy of food industries. Chemical treatment is required essentially for waste released from metal plate industries.

Utility of bacteria is restricted in effluent treatment due to harmful chemicals. Generally, purification of industrial effluents involves sedimentation, coagulation, microfiltrate, adsorption and electrochemical methods.

5.2.1.1 *Toxic metals:* The most toxic metals present in industrial effluents are heavy metals. There are extremely harmful to plants and animals including human beings. Some of the harmful heavy metals are mercury, lead, cadmium and arsenic. There are three important steps involved in effluent treatment—(i) precipitation (ii) cation exchange and (iii) solvent extraction.

5.2.1.1.1 (i) *Precipitation:* Precipitation involves a reagent that can produce an insoluble compound with the metal. To precipitate noxious metal, excess of anion is added for ex: to precipitate Hg_2^{+2} ion from solution, excess of chloride ion would be used.

5.2.1.1.2 (ii) *Cation exchange:* Cation exchange is ideal an choice to remove heavy metal ions like copper, lead, cadmium etc. Ion-exchange coupled with precipitate methods work efficiently in most cases.

5.2.1.1.3 (iii) *Solvent extraction:* Several organic solutions used to breakdown noxious materials into carbondioxide, water and oxides of sulphur etc. Electrochemical method is often used to remove oil and paints in aqueous system. Electrolysis along with ion exchange is frequently used to remove dissolved salts present in industrial effluents.

5.3 SEWAGE TREATMENT

Sewage contains industrial effluents, organic materials like human waste, food waste and cleansing compounds. Sewage contains both pathogenic and non pathogenic microorganisms.

Table 5.1. Types of heavy metals and their effect on human health

Pollutants	Major sources	Effect on human health	Permissible level
Arsenic	Pesticides, fungicides, metal smelters	Bronchitis, dermatitis	0.02 ppm
Cadmium	Welding, electroplating, pesticide fertiliser, Cd Ni batteries, nuclear fission plant	kidney damage, bronchitis, gastrointestinal disorder, bone marrow cancer	0.06 ppm
Lead	Paint, pesticide, smoking, automobile emission, mining, burning of coal	Liver, kidney, gastro-intestinal damage, mental retardation in children	0.1 ppm
Manganese	Welding, fuel addition, ferromanganese production	Inhalonation or contact causes poisoning of central nervous system	0.26 ppm
Mercury	Pesticides, batteries, paper industry, fluorescent tubes etc.	Damage to nervous system, protoplasm poisoning, kills fishes and fish eaters	0.01 ppm
Zinc	Refineries, brass manufacture, metal plating, plumbing	Zinc fumes have corrosive effect on skin, cause damage to nervous membrane	15 ppm

The main components of sewage are 99.5% of water and 0.1-1% inorganic and organic matter in suspended soluble form. It also has higher BOD value. The Biological Oxygen Demand (BOD) designates the amount of degradable organic material present in the water sample. Decrease in BOD of waste water during treatment reflects the effectiveness of treatment in transforming organic to inorganic waste. In addition, sewage contains suspended mineral matter, bacteria, viruses and dissolved salts. Sewage treatment is classified into three types—primary, secondary and tertiary treatment (Fig. 5.3).

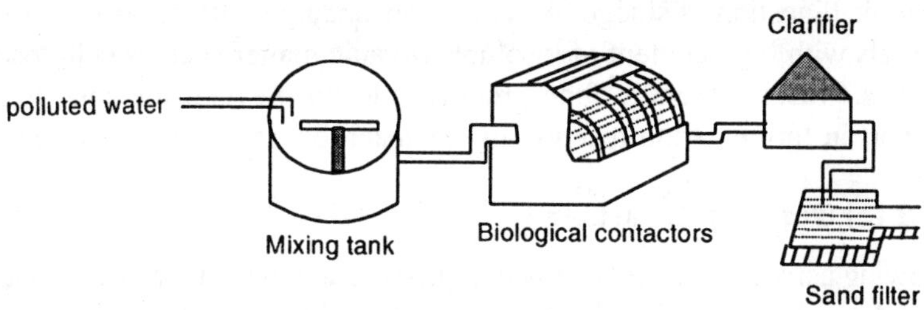

Fig. 5.3. Microbiological treatment of polluted water.

5.3.1 (a) Primary treatment

Primary treatment consist of removal of large objects such as paper, rags, plastic and pieces of wood by passing the waste water through a series of screens. These are then allowed to settle for a period of atleast 90 minutes to 2 hr. At this stage chemical is added to settle down flocculate particles so that they undergo sedimentation rapidly.

5.3.2 (b) Secondary treatment

Secondary treatment of sewage is designed to stabilize organic materials and reduce BOD in the sewage. In secondary treatment, microorganisms like bacteria and fungi are used to biodegrade organic materials present in the supernatant of the sedimentation tank. Microbial oxidation takes place in an aerated tank. The sediment is transferred to anaerobic digestion tank for degradation. After microbial oxidation, effluents are then moved to second sedimentation tank. From there sediment is shifted to anaerobic digestion tank. During this treatment sewage provides nutrients for anaerobic organisms and facilitated the growth. Different microbes degrade larger organic molecules into smaller ones and then are further oxidised by other organisms. After the treatment BOD is generally reduced to 75%. During secondary treatment most of the pathogenic bacteria get destroyed.

5.3.2.1 *Trickling filter tank:* It is 3-10 feet deep tank packed with either crushed rock or some other inert support material, above which micro-organisms forms a thin film. In secondary treatment the liquid waste is applied evenly to the top of the filter and it slowly percolates downward, depositing organic matter and the microbial film on the support. The special feature of trickling filter is aerobic conditions, maintained by an upward flow of air through the spaces present between the particles of solid support. Effluents from the filter tank is transferred to sedimentation tank where suspended solids are allowed to settle. The efficiency of trickling filter, depends primarily on the contact area and time i.e. size and shape of the particle in the solid support material.

5.3.2.2 *Anaerobic secondary treatment:* Aerobic digestion is followed by anaerobic digestion of sludge. Anaerobic secondary treatment is used to treat materials with high content of insoluble organic matter such as cellulose and industrial waste especially food process industry. Several methane bacteria take part in fermentation reaction in the anaerobic secondary treatment.

5.3.3 (c) Tertiary treatment

Following activated sludge treatment, lime is added to treat water for coagulating and precipitating phosphate containing particles. These are then settled in clarifier. Some of the filters, made of coal and sand remove remaining particulars. Activated charcoal removes detergent, pesticides and other toxic materials. Chlorine is added to purify water in the final step to destroy remaining microorganisms. Further, chemical precipitation removes phosphates and nitrates (Fig. 5.4).

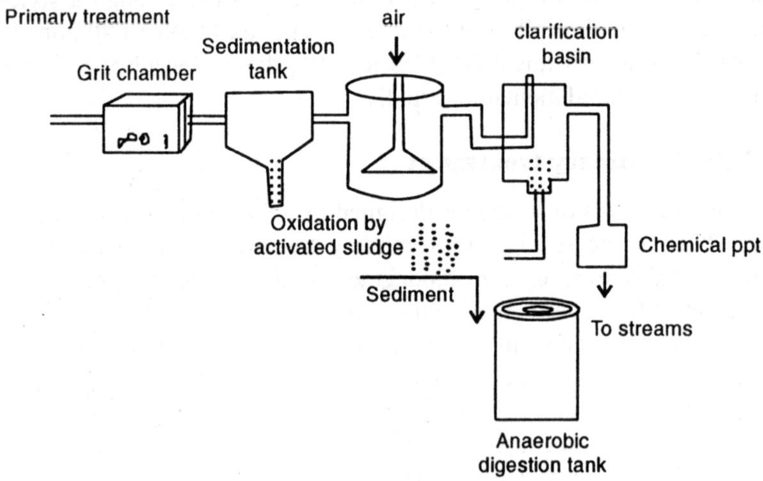

Fig. 5.4. Sewage treatment.

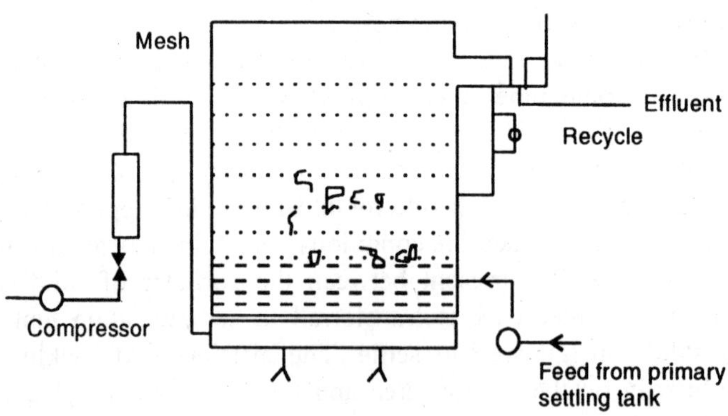

Fig. 5.5. Sludge process.

5.3.4 Immobilisation Technique in Waste Water Treatment

The immobilisation of cells and enzymes are used in waste water treatment to reduce BOD, detoxify pollutants and bio conversion of waste to get specific product. Floculation of cells that occurs in activated sludge process may be considered as crude or preliminary way of using immobilised enzyme.

Use of immobilisation amylase enzyme is suggested for treatment, waste-water from the wheat and starch industry. In one approach *Micrococcus denitrificans* cells can be encapsulated for reduction of nitrate to nitrite. Nitrisomonas cells immobilised in alginate have been used to oxidise ammonia in waste water to NO_2 and NO_3 which prevents algae growth.

QUESTIONS

1. Briefly outline the treatment of municipal waste and industrial effluents. (BU Nov. 1999)
2. Explain the methods of treating industrial effluents. (BU May 1996)

SHORT QUESTIONS

1. Composting
2. Vermicomposting
3. Effluent treatment
4. Types of waste
5. Trickling filters
6. Tertiary treatment
7. Immobilisation technique
8. Pasteurisation of milk

6

Biodegradation

Biodegradation is a process in which microorganisms are committed to transform toxic chemicals into less toxic or harmless. The organic compounds are often classified as biodegradable, persistent or recalcitrant. A biodegradable organic compound is one that undergoes a biological transformation; a persistent organic compound does not undergo biodegradation in certain environment and finally recalcitrant compound does not undergo biodegradation in any condition.

The microorganisms can specifically act on toxic waste by secreting of enzymes. Microorganisms excel at using organic substances as sources of nutrient and energy. These include detergents, triculoroethane, xylene, toluene etc. The petroleum consists of complex array of alkenes, cycloparaffins, aromatic compounds and sulphur compounds. About metric tonnes of petrochemicals are transported into the oceans and surface of the soil and later fed on by microorganisms. It is estimated that 10% of global oil deposits was destroyed in this manner. Coal is another reserve of highly reduced carbon compounds. Many microorganisms can utilise the carbon compounds in coal as substrates and can metabolise the sulphur compounds. Many organohalogen compound such as polychlorobiphenyl (PCB), dichloro-diphenyl trichloroethane (DDT) are widespread and long-lived human made contaminants in soil. It was discovered that the microorganisms are able to degrade all halogenated synthetic organic compounds.

6.1 FACTORS AFFECTING BIODEGRADATION

Several factors are responsible for biodegradation of residual compounds. Some of the factors like presence of substitution, solubility can play an important role in biodegradation etc. Substitutes like polar groups provide site for microbial attack. In addition, temperature, heavy metal concentration play an important role in biodegradation process.

6.1.1 Effect of Pesticide

Pesticides can interfere with respiration. Arsenic containing pesticide would interfere pyruvate dehydrogenase enzyme which converts pyruvate to acetyl CoA. The nervous system is greatly damaged by DDT molecules trapped in the pores of the nervous membranes thus distorts the membrane by sodium leaks and diapolarising the nerve cell. These are also carcinogenic and mutagenic, for example 2, 4, 5, T dioxins are highly mutagenic and affect reproductive system.

6.2 BIODEGRADATION OF PESTICIDES

Insects have been a major competitor for food. The farmer incurs huge loss due to insect menace. They destroy agriculture crops, stored products etc. Pesticides are chemicals which effectively combat insect problems by destroying them. Human population is benefitted from the pesticides due to its toxicity against insects.

Modern agricultural practice emphasises on the extensive use of chemicals specially designed to target and destroy insects and pests. Every year large quantities of pesticides are manufactured for agricultural practice. Accumulation of pesticide residue in the soil and direct entry into the environment was noticed recently. A severe environmental problem may result from excessive use of fluorochloro carbons and other pesticidal residue. In terms of worldwide use of pesticides Japan and USA uses 1200 and 3000 grams per hectare respectively. India uses comparatively less of pesticides i.e., 400 grams per hectare. Some hazardous pesticides thrive for more than two decades in soil e.g., DDT, Aldrin and chloradane persist for more than 15 years in soil. Dicamba can be in the soil for more than 4 years.

6.2.1 Classification of Pesticide

Pesticides are classified broadly into (a) organochlorine compounds, (b) carbamic acid derivatives and (c) organophosphorous compounds.

Table 6.1. Pesticide group

Insecticide	Organochloro pesticide	Herbicide	Fungicide
Deltoc merthrin	Altrazine	Sulfometuron	Triazole
Chloropyriphos	Aldrin	Round up	
Cypermerthrin	DDT		
Fenvalerate	BHC	*Auxin herbicide*	
	Endosulfan	-2,4-D	
		Dicamba	
		Pichloram	

Organochlorine compounds are polychloroterpenes and polychloro-cytodions. Derivatives of phosphoric acid and phosphotioicacid placed under organo phosphorus compounds. Pesticides are insecticides, herbicides, nematocides and rodenticides. (See Table 6.1.)

6.2.2 General Mechanism of Biodegradation

The degradation process of pesticides are as follows:

(a) Substituent on aromatic ring of pesticides are removed.
(b) Aliphatic carbon chains are degraded by oxidation process.
(c) Aromatic rings are degraded by oxidation process.
(d) It substitutes are methyl, carbonyl groups, they can be removed from blocking positions metabolically.

Biotechnology holds promising role in reducing pesticide problem especially in heavily contaminated soil. The advancement in genetic engineering offers attractive and significant solution to pesticide waste disposal. The genes for biodegradative enzymes can be engineered to enhance efficient pesticide degradation. Several fungi and bacteria are used to degrade pesticide.

6.2.3 Microorganisms for Biodegradation

Biodegradation of pesticide is done by certain useful microorganisms. They are used in cleaning up of soil containing 20,000 ppm of 2,4-T. Plants can even grow in microbial treated soil. Several microorganisms like *Pseudomonas, Arthobacter* are capable of degrading phenoxy acid. Fungi like *Aspergillus, Rhizopus, Fusarium* are capable of degrading certain herbicides like atrazine. Similarly aliphatic acids are degraded by bacillus, *Pseudomonas* and *Agrobacterium*. The yeast and *Neurospora* are well known for their potential to degrade thiocarbonate and chloroneb respectively.

6.2.4 Biodegradation of DDT and Aldrin

DDT is generally degraded under anaerobic conditions. Certain bacillus species *Clostridium, Pseudomonas, Streptococcus* contribute significantly in DDT degradation. The insecticide aldrin is converted to dieldirin.

Biodegradation of Coumphos

Flavobacterium species can rapidly degrade organophosphates insecticide by enzyme parathion hydrolase where insecticide coumphos is converted to less toxic products.

6.2.5 Steps in Biodegradation of Some Pesticide

6.2.5.1 (a) *Parathion:* This organophosphorous insecticide undergoes a very rapid hydrolysis in the environment. The nitrogroup of parathion is reduced by soil microorganisms to give non-toxic aminoparathion, which is stable in the soil. At this juncture it is not clear whether inactive parathion creates environmental problems.

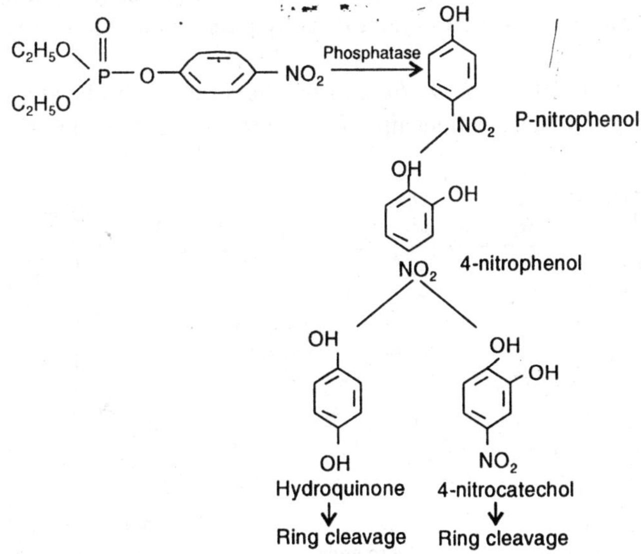

Fig. 6.1. Biodegradation of parathion.

6.2.5.2 (b) *Auxin Herbicide 2,4-D:* 2,4-D is used as a weedicide and is subjected to rapid environmental degradation. These compounds are cleaved by microorganisms to the corresponding phenol which are readily degraded further.

Table 6.2. Flow chart showing pesticide entry into Environment

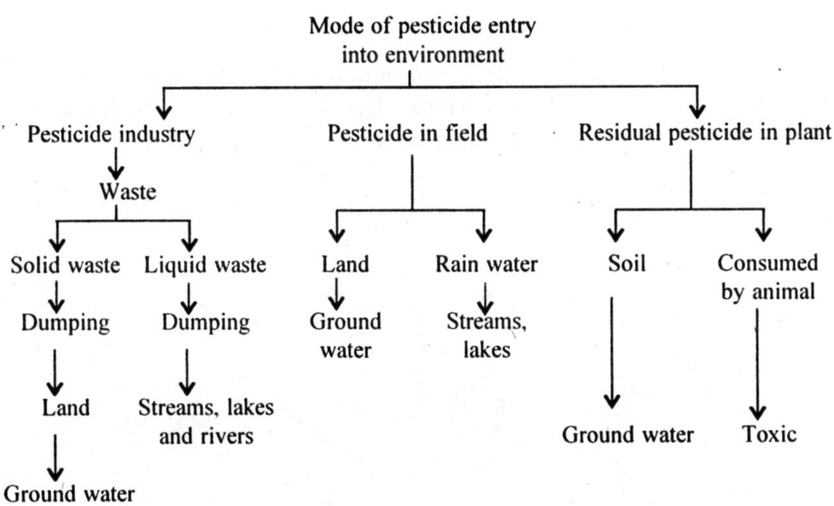

6.2.6 Biodegradation by Immobilisation

Immobilisation is a novel technique in which cells, microorganisms and enzymes are immobilised or entrapped or imprisoned in distinct phase. Several pesticides can be converted to nontoxins by passing through entrapped cells or enzymes.

6.2.6.1 Methods of Immobilisation: Several methods are available depending upon the relationship of the enzyme or cells to the polymer matrix.

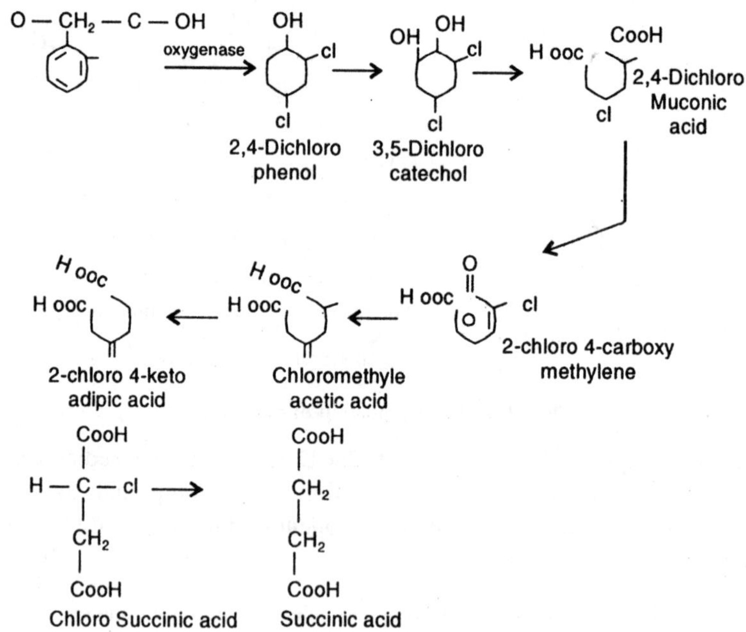

Fig. 6.2. Biodegradation of 2,4-D.

6.2.6.1.1 *Adsorption:* Involves adsorption of cells or enzymes to polymer matrix, perhaps one of the most widely used techniques in which enzyme catalysts are electrostatically bonding to an ion exchange cellulose.

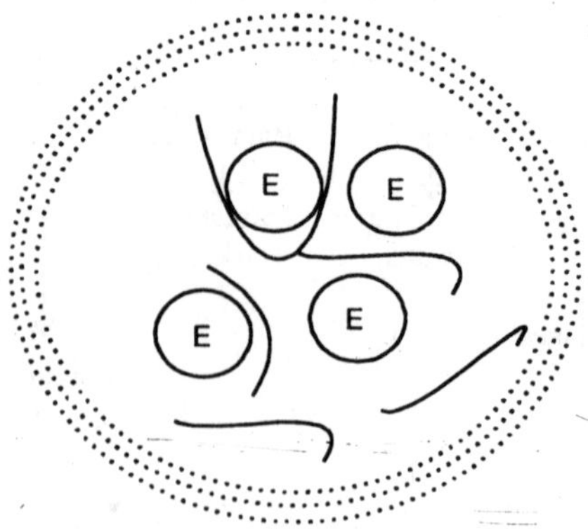

6.2.6.1.2 *Covalent bonding:* Covalent method is specific for enzyme immobilisation. Principally there are two ways of covalent bonding of an enzyme to a polymer. First form involves activating the polymer with a reaction group. The second in the use of reagent to link or bridge between enzyme and polymer.

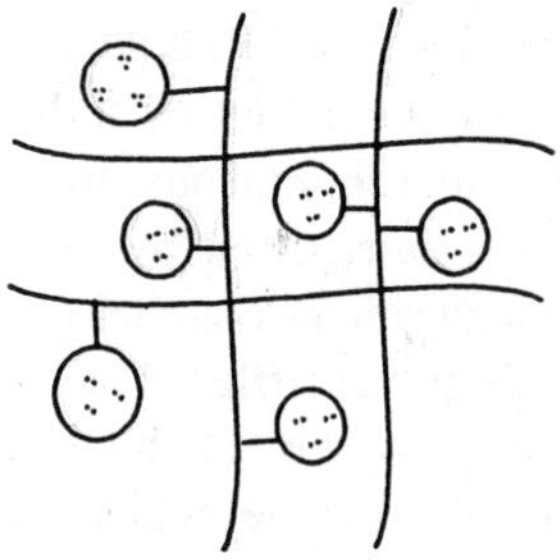

6.2.6.1.3 *Encapsulation and Entrapment:* Encapsulation technique is simple and economically cheapest. It is completely non toxic to enzymes. Encapsulation involves cells, whole microorganisms, enzymes in a alginate solution is dropped in calcium chloride to form a bead or droplet. The cells or enzymes get encapsulated within the beads.

Entraparent of enzyme within matrix is easy to perform. Two types of polymers are found to be applicable. One is agar or alginate and the other is Polyacrylamide gel.

Immobilisation of parathion hydrolase used for waste water containing pesticides have been studied. These enzymes can be used to detoxify pesticide effluent. The whole fungi or enzyme is immobilised on glass beads packed inside the column through which effluent is allowed to pass down. Immobilised enzyme can act on it and knock down the toxicity of pesticide. The enzyme can be reused to detoxify pesticides. The acrylamidase enzyme of *Fusarium solani* was imprisioned to degrade propanil into nontoxic 3,4-dicholoroaniline.

6.3 BIODEGRADATION OF OTHER TOXIC CHEMICALS

Degradation of explosives

The waste stream from the explosive industry are often discharged to water, soil and lakes. The waste stream contains 2,4-6, trinitrotoluene (TNT), hexahydro 1,3,5 trinitro 1,3,5 triazins (RDX) and N-methyl 1,2,4-6 tetra nitroaniline (Tetryl). Since they are sparingly soluble in water, they have largely remained in the soil, especially at discharge site. Atleast $ 300 per ton is required to decontaminate the soil. The cost would exceed billions of dollars in reality. Thus considerable endeavour is given to develop cheaper methods. Studies have shown that RDX and HMX can be degraded by anaerobic bacteria and the TNT can be biodegraded under both aerobic and anaerobic conditions. Biodegradation of over 90% of explosives was achieved within 80 days in a compost pit maintained at 55°C. After 150 days, the initial concentration of 18,000 mg of explosives per kilogram of soil was reduced to 74 mg/kg.

6.4 BIODEGRADATION OF PENTACHLOROPHENOL

The compound pentachlorophenol (PCP) is generally toxic to living organisms. The global production of PCP is about 60 million kg/yr and is proved as a potential killer of weeds, microorganisms and molluscs etc. The study on biodegradation of PCP was carried out by the US Environment protection agency, Minnesota. The experiment channel consists of ponds with mud at the bottom alternating with gravel riffles. They are fed from a natural source of water that contains diverse microbial habitats. Rapid microbial degradation of PCP in the treated channel became evident in about 3 weeks.

6.5 BIODEGRADATION OF BENZENE AND OTHER AROMATIC HYDROCARBONS BY *PSEUDOMONAS*

Benzene is degraded by bacteria in two steps. The initial dioxygenase mediated hydroxylation is followed by dehydrogenation. The enzyme dioxygenase added two atoms of oxygen into benzene ring resulted in altered benzene structure, i.e. 1,2 dihydroxyl 1,2 dehydrobenzene. It is then converted to catechol by enzyme benzene glycol dehydrogenase. These are two common pathways for the degradation of other aromatic hydrocarbons. The subsequent steps in metabolism of catechol is either oxidised by catechol dioxygenase and cleavage to hydroxy muconic semialdehyde.

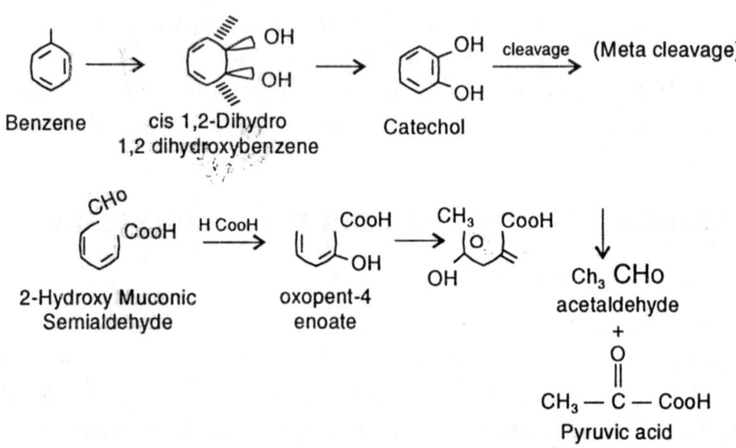

Fig. 6.3. Benzene degradation by *Pseudomonas putida.*

6.5.1.1 *TOL Catabolic Plasmid (PWWO):* The *Pseudomonas putida* (Mt[2]) containing plasmid is known as PWWO. This plasmid encodes genes for metacleavage pathway of hydrocarbon degradation. The TOL plasmid, PWWO has shown to confer toluene degradation and other benzene derivatives. This plasmid TOL containing several genes encoding for catabolic enzymes i.e. xyl genes. These genes are organised into two operons responsible for degradation by upper and lower pathway. The upper pathway contains the genes, xyl CAB encodes the enzyme helps in degradation of toluene, xylene, benzoate and toluates (methyl benzoates), respectively. The lower pathway contains genes encodes the degradation of benzoate and toluates to acetaldehyde and pyruvate.

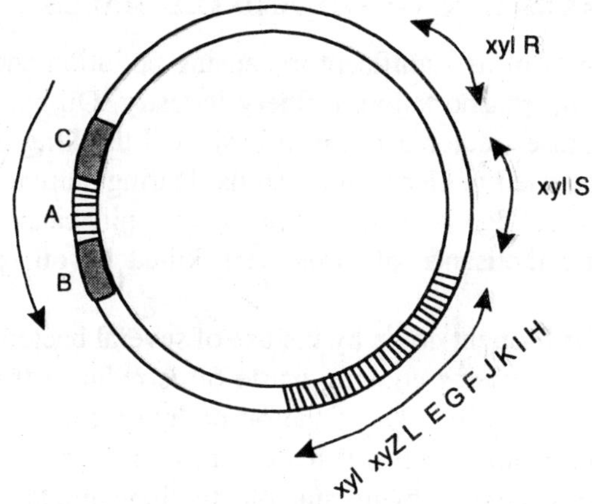

Fig. 6.4. TOL (PWWO) catabolic plasmid.

When a toluene enters the cell, it binds to xylR proteins to form xylR-toluene complex. This complex binds to the promoter of the xyl CAB operon and initiates transcription by influencing the binding of RNA polymerase. In lower pathway, the benzoate (product of toluene degradation) binds to xylS protein and this complex binds to xylS-protein and activates transformation of the genes. The degradation of toluene by lower pathway is organised in plasmid as xyl x y z L E G F J K I H (Table 6.3).

Table 6.3. Types of genes in TOL plasmid of *Pseudomonas putida*

Genes in upper pathway	Genes in lower pathway
xyl A	xyl x, y, z
xyl B	xyl E
xyl C	xyl F
	xyl G
	xyl H'
	xyl I
	xyl J
	xyl K
	xyl L

6.6 BIODEGRADATION OF PHENOLS

Phenols are recalcitrant pollutants are toxic to animals when exceeds 4mg/l of water. Excess of phenols in water can increase BOD. Several targeted microorganisms are employed in bio-treatment to degrade phenols. Spores of *Fusarium flocciferum* can be used to degrade phenol by immobilisation process. This process can tolerate high phenol content.

6.7 BIODEGRADATION OF OIL AND OIL SPILLS

The contribution of oil is significant in causing pollution and it is produced during production, transportation, refinery industry. Oil can accumulate on water and this causes major damage to birds and drinking purpose. Microorganisms can be used to clean up oil spills. During shipment of petroleum products due to leakage or ship wreckage oil spill causes serious water pollution. Several thousands of birds were killed by oil spills during the Gulf war.

Oil spill can be brought down by the use of several bacterial mutants. The activated sludge containing cooking waste oil is reduced from 250 to 2mgl by mutant bacteria. Nearly 60% of the oil pollution can be removed within 72 hours by biotreatment. Several hydrocarbon degrading microorganisms are equipped with target enzyme suitable for biodegradation of oil spills. *Pseudomonas* strains contain OCT-octane, NAH-napthalene and TOL-Toluene genes. The plasmid containing several genes coding for enzymes, which are targetted on systemic degradation of the compounds. The first patent was issued for genetically engineered *pseudomonas* bacterium, which was engineered by Indian born American scientist Anand Chakraborthy. The genetically engineered pseudomonas was found to degrade oil spill effectively.

6.8 BIODEGRADATION OF PLASTICS

Plastic is non-biodegradable. It is made up of chemicals like polypropinyl chloride, polystrene, and polyethylene. The pollution from plastic is mainly due to its non biodegradable nature. Due to persistent effort in research many biodegradable and non biodegradable plastic have been developed by a Japanese Company. Biodegradable plastic was designed based on copolymerisation of two different aliphatic polyster derivative.

Biodegradable plastic contains poly B hydroxybutyrate (PHB) which are synthesised by microorganisms like a *Alkaligenes, Pseudomonas* which act as an energy reserve material for microbes. The extracted PHB from microorganisms are very useful in making biodegradable plastics. The direct moulding and extrusive of dried cells is possible, under this PHB are high which is very useful biodegradable plastics.

In *Alkaligenes* eutrophic biodegradable poly (3-hydroxybutyrate) i.e. PHB is synthesised from fructose or butyrate. These key enzymes in three reactions steps are involved in PHB synthesis. The biochemical pathway leading to poly (3-4 hydroxybutyrate) production from the metabolism of acetyl CoA, which is then transformed to (poly-3 hydroxybutyrate) by a three step reactions. These are catalysed by a keto thiolase NADPH-linked aceto acetyl-CoA reductase by a poly (3-hydroxyalkanoate) polymerase respectively.

6.9 BIODEGRADATION OF CELLULOSE

Several bacteria and fungi are commonly employed in the biodegradation of cellulose, a major component of plant cell wall and the whole biomass. Extensive description on microbial degradation of cellulose is described in the third chapter.

QUESTIONS

1. Give an account of Biodegradation of pesticides. (BU April 2001)
2. What is Biodegradation? Explain the role of microorganisms in biodegradation.
3. Explain the process of biodegradation of toxic chemicals.

SHORT QUESTIONS

1. Biodegradation of plastics
2. Biodegradation of oil and oil spill
3. Biodegradation of Hydrocarbons
4. Biodegradation of Herbicides
5. Biodegradation of pesticides (BU Nov. 1999)

7

Biological Control of Insects (Biopesticides)

Insects are the largest group of animals which thrive well in all kinds of geographical conditions on the planet. Rapid adaptibility of insects to new environment is amazing. Insects are the major menace to our crops, resulting in huge loss and negligible crop yield. Hence, protection of crop from insect menace is indispensable and several biotechnological strategies have to be evolved to combat these problems for several years. When DDT was introduced to control insects, they acquired resistance immediately. Besides DDT caused serious problem to public health and this forced the government to ban the use of DDT. Several chemical insecticides like pyrithrin, organophosphorus have also been introduced.

Several microorganisms act as natural enemies to insects. We can use these beneficial microorganisms as a powerful biological weapon to kill or control insects. Several microorganisms like, bacteria, virus, fungi and even protozoa can be effectively marshalled to fight insects. Hence, biological control of insects is one of the most effective method to protect our plants. Biological insecticides have the distinct advantage of being biodegradable. They do not release toxic residues that contaminate environment.

7.1 BACTERIAL INSECTICIDES

Bacterial pathogens of insects are generally classified into spore forming and non-spore forming forms. These two forms are highly potential toxins against wide variety of insects. Some of the well known bacterial insecticides are *Bacillus thuringiensis, Bacillus popillae, Clostridium brewigacions* etc.

7.1.1 Thuringiensis toxins as a natural pesticide

Of all the biological agents that have been evaluated, the most effective and successful one is the bacterial insects pathogen, *Bacillus thuringiensis* popularly known as BT.

Bacillus thuringiensis is a spore forming, gram positive crystalliferous bacteria that produce proteins that are insecticidal. The proteins in BT are found as crystalline bodies with sporulating bacteria.

B. thuringiensis sub. sp. *kyrstaki* was first discovered in 1902, but its significance as an insecticide was realized after 1951. The insecticidal crystal proteins are characterised by their potency and specificity towards specific pest. They are safety to non target insects, vertebrates particularly humans.

Bacillus thuringiensis (BT) have enjoyed a long history of use in horticulture industries, where the mixture of crystals and spores of BT are sprayed just like a chemical pesticide but they have not been used with much success on large acre field crops. BT are of two types. One is very effective against lepidoptaran insect pest, another is against a number of dipteran flies. The crystal protein of BT is of single polypeptide of 130 kd while the toxic peptide is half that size. This indicates that only half the portion of this protein is actually an insecticide active peptide. The genes encoding crystal protein are present in plasmids rather than in chromosome of bacteria, for example, in HD-1 strain, there are two plasmids in bacteria containing insecticidal toxin gene. Several factors influence the insecticidal activity of strains. They are number of toxin genes and qualitative levels of expression of various genes.

7.1.1.2 *Mode of Action of Insecticidal Crystal protein:* The insecticidal crystals are composed of large protein that is essentially inactive when a caterpillar injest some of the insecticidal crystal proteins. The alkaline conditions of the insect midgut, plt 7.5 to 8.0 cause the crystal to dissociate and release the crystal protein. At this stage crystal protein is inactive, but the presence of possible specific proteases within the gastric juices of the insects breakdown to its protease resistant core and convert the inactive protein to become active protein immediately. This activated insecticidal protein then binds to a specific receptor on the border membranes of the cells lining the midgut and insect itself into the cells membranes, when about eight of these aggregate together, this form a pore or channel through the membrane, and allow the cell content to leak out continuously causing the death of the cell. These are very essentially participate in nutrient absorption. The insect stop feeding immediately and finally starve to death possibly within 24 hrs. The process of crystal solubilisation and proteolytic processing to an active insecticidal protein and binding to a specific receptor, all make the BT protein a highly specific potential insecticide.

7.1.1.3 *Isolation of Toxin Gene:* During isolation of toxin gene it is essential to confirm whether the toxin genes are located on a plasmid or on the chromosomal DNA. To confirm the presence of toxin gene on plasmid BT strain can be conjugated with a strain that lacks insecticidal activity. *Bacillus thuringiensis* strains are cultured and cells are lysid. The total DNA is isolate and separate plasmid and chromosomal DNA fractions by caesium chloride gradient centrifugation. When toxin gene is encoded in plasmid it is then subjected for sucrose density gradient centrifugation. The plasmid DNA sample is then subjected for fractionation. The medium and larger plasmids are treated with restriction enzymes and ligated into PBR 322 plasmid. The clone banks were transformed into E Coli and then screened by immunological method.

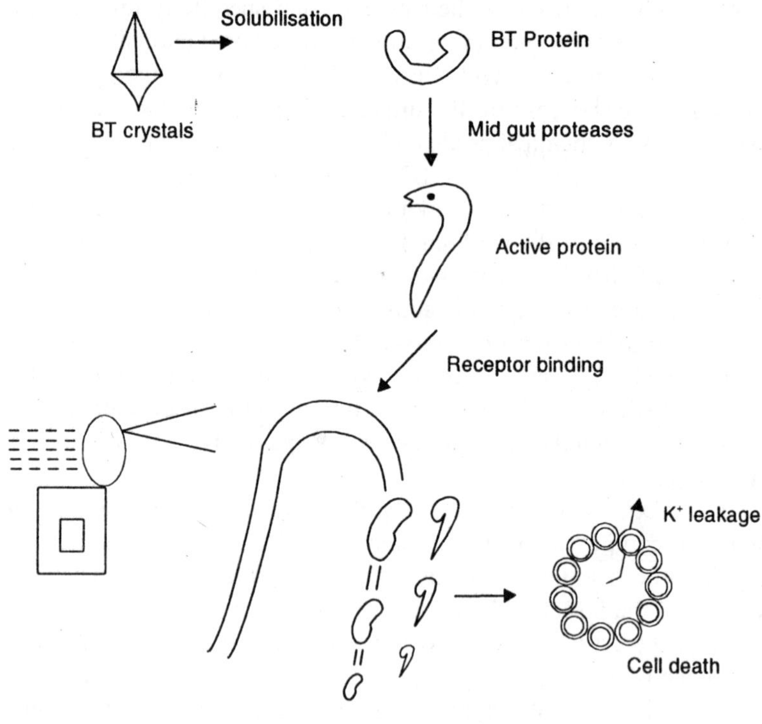

Fig. 7.1. Action of BT toxin protein.

7.1.1.3 *Classification of crystal protein:* Many thousand different types of B-thuringiensis have been collected and their insecticidal protein content and activity is determined. A large number of BT-insecticidal protein gene have also now been cloned and sequenced. The researcher have proposed naming system for its crystal protein gene. Commonly known as cry genes, based on their protein sequence. The protein encoded by cry I genes are toxic only to caterpillar. The cry genes encode protein toxic lepidoptera or diptera (flies and mosquitoes) while the cry IV proteins are only active against Diptera. However, cry III genes produce protein active against beetle i.e. coleoptera larvae. In the classification even smaller division have been made. The cry I group, for example was divided into cry IA, IB, IC. Even fine sub-division have also been made cry IA consist of Cry IA (a) IA (b) and cry IA (C) (Table 7.1).

Table 7.1. Classification BT crystal gene

Crystal type	Insect Host	Size (kD)
Cry I A	Caterpillar	137
Cry I B	Caterpillar	138
Cry I C	Caterpillar	134.8
Cry II A	Caterpillar	70.9
Cry II B	Fly larvae	70.8
Cry III A	Beetle larvae	73.0
Cry IV A	Fly larvae	134.4
Cry IV B	Fly larvae	127.8
Cry IV C	Fly larvae	77
Cry IV D	Fly larvae	72.4

7.1.1.4 *Commercial product of BT:* The most widely used BT insecticides are *Bacillus thuringiensis* var. Kurstaki. These isolate are highly toxic only to butterflies and moths. The US Environmental protection agency documented several BT product. Several BT formulations have already been in commercial use by different trade name such as Thuricide, Diphel, Javelin, Bactospeine and SoK-BT. Many small companies commercialised BT product and widely available as wettable powder, liquid concentrate and granules. They used to control variety of insects.

Table 7.2. Bt formulation that control caterpillar

Bag worms
Tent caterpillars
Gypsy moth
Forest caterpillars
European corn borer
Fall web worm
Spruce bud worm

Several other Bt formulations began in early 1980 to target black fly and mosquito (Table 7.2). The Bt products are obtained from *Bacillus thuringionsis.* var Israelis. In 1988, *Bacillus thuringiensis* var Sandiago sold under the trade name M-one was first registered for use against larvae of the colorado potato bicycle. This product also kills leaf beetle adults and larvae. In addition, the registration of another product Trident, contains *B thuringiensis* var. tenebrionis.

7.1.1.5 *Improvement of BT through Genetic Engineering:* One of the limitations and concern with use of BT is attributes on human safety and safety to non target organisms. The main objectives of BT improvement is to ensure construction of fusion protein in which BT with different host spectra are fused together. Direct modifications of the specificity domains in the proteins.

Almost all the BT crystal protein contains five regions of conserved amino acids. These regions are perhaps responsible for toxicity. Similarly, almost all Bt protein contains regions in which amino acids are highly variable. The variable region is believed to be responsible for different host spectrum. It is possible to modify this variable region through RDT in order to increase host range. Another strategy is to fuse Bt toxin domains to proteins capable of binding to specific sites in midgut cells.

7.1.2 Other Bacterial Insecticides

7.1.2.1 *Vegetable Insecticidal Proteins* (VIPS): Insecticidal activity have been observed by fluids collected from *Bacillus cereus*. These vegetable insecticidal proteins (VIPS) are very active and potential toxin to northern and western corn root worms. Cloning and characterisation of the gene encoding this activity yielded a novel insecticides protein. ViP3A displayed insecticidal activity against wide spectrum of Lepidaptera worms. Studies showed that when insects were fed on diet supplemented with ViP3 complete lysis of midgut epithelial cells followed by larvae death.

7.1.3 Cholesterol Oxidases

Cholesterol is required for structural integrity and normal functions of all the cell membrane. Any process interfering cholesterol synthesis and availability would disturb cellular integrity. Interestingly two *streptomyces* culture filtrate contain an enzyme cholesterol oxidases (CO). A protein showed acute toxicity to boll weevil larvae. When insect injested these proteins toxicity occurs at midgut epithelial cells accompanied by local cystolysis. Suggesting cholesterol oxidases to be responsible for alterations in the membrane cholesterol.

7.1.3.4 *Bacillus sphaericus: B. sphaericus* an aerobic spore forming bacterium acts potential insecticide against certain mosquito species. These are more active than *B. thuringiensis*. The dry powder preparation of *B. spaericus* when sprayed in uncleaned places which provided control of mosquito larvae. *B. spaericus* can acts as a potential larvicide worldwide in future.

7.1.3.5 *Bacillus Popilliae: B. popilliae* is a pathogen of beetles. this bacterium does not produce toxins instead it multiplies rapidly when injested by the larvae. This bacterial species invade the hemocoel multiply rapidly in 72 hrs. and killing the larvae in this process. The commercial preparation of spore have been done in Japan by growing bacteria in Japanese beetle larvae.

7.2 FUNGAL INSECTICIDES

Fungal insecticides are otherwise known as Mycoinsecticide which are potential against insects. More than 200 species of fungus can infect insects. The most widely used fungus *Beauveria bassiane* is well known for its efficiency in controlling more than 500 insects. Another fungus *metarrhizium anisopilae* popularly known as muscardine fungus can infect at least 200 insects. In Europe and USA the two fungus *Beauveria bassiana* and *metarrhizium anisopilae* have been widely used to control certain bugs and beetles. These two fungal formutations are commercially sold as Boverin and Metagnine. Mycer is commercially available fungus Hirsutella thompsonis in USA to control citrus rest mite. *Verticillum lecanii* is produced commercially in U.K. for marshalling against aphids. Some of the fungal formulations are prepared and sold as Dust or Wettable powder. At several instance mixing DDT with spores of *Beanvaria Bassiana* was found to be effective in control of insects. Recently a fungus parasite *parcilomyles farinosus* has been reported on white fly *Benissia fabaci.*

Mechanism of action of mycoinsecticide: In the infection process the fungal conidium binds to the insect cuticle.

The germinated tube once enters ramifies and kills insect by continuous toxic metabolites. Although industrial production of entomopathogen fungus not yet materialised. Its role as proved highly promising in pest control.

7.3 VIRAL INSECTICIDE

Viruses have undergone extensive evaluation and have been registered for use against insects. Due to their specificity and effectiveness viruses have greatest potential for use in the control of integrated pest management.

Among several types of insect viruses the most promising over which lethal against economically important viruses are nuclear polyhedro virus (NPU) and Granulosis Virus (GV). Insect viruses belongs to the family Baculoviridae. Among viral insecticides occluded virus are beneficial as they are made up of proteinaneous and silicaceous cover. Which are resistant against harsh environmental factors such as temperature, UV radiation etc. Approximately 280 insect viruses have been described. Nuclear polyhero viruses are rod-shaped viral particles embedded within distinct polyhedra particles embedded within distinct polyhedra inclusion bodies. These inclusion bodies are 0.5-3.0 micrometer and is visible under microscope. To nucleo capsids or virions in each polyhedron is made up of double-stranded DNA measure upto 430 μm in length. Majority of the NPV viruses are narrow range, insecticidal activity. However, some isolates from alfa alfa caterpillar *Autographa californica* have a broad spectrum insecticidal activity, shows target upto 40 lipidoptera species (Table 7.3).

Since virus requires living media to grow, large scale production is still in infant stages. The need of larvae for NPV virus culture have registered in USA. Currently several European countries like Denmark, Sweden, USA, Australia have already been implemented practice of using viral insecticides. In India, some promising work have been down in the use of viruses on sun flower, cotton and fruit species like tomato.

7.3.1 Mechanism of Action

Once viruses are injested by insect, it multiply rapidly, proliferate and enters cytoplasm and nucleus finally knock down the host. The injested virus in the midgut regions undergo disintegration due to alkaloid condition. The released virions or nucleic acid is then travel from gut wall into the body cavity. The cytoplasm and nucleus of the cells are targetted by virus rods results in extensive lysis as the infection advances. The insects colour changes rapidly from whitish to pinkish color.

Table 7.3. Baculovirus as Viricides

I.	*Nuclear polyhedro virus*	*Target insect*
	Anticarsia gemmatalis NPU	*Anticarsia gemmatalis*
	Autographa californica NPU	*Orgyia pseudotsuga*
	Gilpinia hercyniae NPU	*Gilpinia* sp.
	Lymantria dispar NPU	*Lumnatria dispar*
	Trichoplusiani NP	*Trichoplusiani*
II.	Granulosa viruses	
	Cydia pomendla Gv	*C. pomonella*
	Plodia interpunctella GV	*P. interpunchella*
III.	Non occluded baculovirus	
	Oryctes rhinocerus virus	*O. rhinocerol*

7.3.2 Inoculum preparation

Viral insecticide when sprayed on the foliage part of the plant. They may be destroyed by environmental factors like U.V. However, viruses can exist and survived for longer period of time in soil. In order to fix viruses on the foliage parts. They are treated with some jaggery, groundnut and cotton seed etc. The viruses can also be fixed in ground net cake and used to control insect.

Table 7.4

Commercial product	Target insect
Virox	*Neodiprion sertifor*
Neo-check	*Neodiprion sertifor*
Elcar	*Heliolhris* sp.
SAN	*Autographa californica*
VPN	*Spodoptera* sp.

7.4 PROTOZOAN INSECTICIDE

Protozoans are unicellular organisms. Several of these are giant killer of insects. Species of microsporidia are the most promising candidates for mass production into microbial insecticide. the microsporidial spores after injestion enters inside gut region and rapidly proliferate by vegetatively in the cytoplasm of the cell. The infected insect develops chronic disease culminate in the death. In USA grass hopper is controlled by protozoan *Nosema locustae.*

7.5 PLANTS AS BIOPESTICIDE

To control environmental problems nurtured by the use of chemical fertiliser have made us to change strategy. Apart from using bacterial, viral and fungal insecticide some of the plant extract can be effectively used as biopesticides to check insects in control. The pesticides of biological origin are safe and environmental friendly. Production of plant based biopesticide is cheaper in production and it can be produced by farmers using simple methods.

The product from Neem (*Azadaricta indica*) plant is among the most promising biopesticide. The beneficial role of neem as an antimicrobial agent was realised in ancient times. Neem contains variety of secondary metabolites like nimbin, meliantriol and azadirictin. Neem has been found to be very effective against grass hopes, white flies, aphids, stem borers etc.

The pyrethrin compound extracted from *Chrysanthemum cinerarifolium* are specifically targetted for household flying insects like housefly, mosquito etc. The extraction from tobacco, known as Ryanin could be a potential insecticide.

Table 7.5. Source of biopesticides

Name	Parts
1. *Lantana camera* (Lantana)	Leaf
2. *Ocimum basilicum* (Tulsi)	Leaf
3. *Azadiracta indica* (Neem)	Seeds, leaf, bark
4. *Chrysanthemum* sp.	Dried flowers
5. *Michalia champaka*	Flowers

QUESTIONS

1. What are biopesticides? Explain their role in agriculture. (BU Nov. 1997)
2. Explain *Bacillus thuringiensis* toxin as a natural pesticide. (BU Nov. 1999)
3. Explain types of viral insecticides, and their role in agriculture.
4. Give an account on bacterial insecticide.
5. Explain the methods involved in biological control of insects.

SHORT QUESTIONS

1. Biodegradation of pesticide (BU Nov. 1999)
2. Plant as biopesticide
3. Fungal insecticide
4. Viral insecticide
5. Baculo virus insecticide
6. BT crystal toxin protein
7. Bacillus thuriengenis

8

Microorganisms in Mineral Recovery (Biomining)

The bacteria involved in the leaching of metal ores are the most remarkable phenomena among life forms. The central role of bacteria in the leaching of copper from inferior ore went unrecognised. The rock eating microorganisms are said to be chemolithotropic. They obtain energy from the oxidation of inorganic substances.

Today bacteria are being deliberately exploited to recover millions of tonnes of copper from billions of tonnes of inferior ores. More than ten percent copper ore recovered in US is from bioleaching process. The use of bioleaching process overcome pollution level caused during biological method of burning of sulphuric fossil fuel and smelting of sulphide minerals. For many years it was thought that the only microorganism which helps in leaching of metal from ore was the rod-shaped bacteria, *Thiobacillus ferr-oxidant*. This bacteria was discovered in 1957 in acid water. In recent years other microorganisms such as *Leptospirilium ferroxidants*, *T. thioxidants* have been discovered. Among these iron-oxidising bacteria, *T. ferroxidants* are the most popular, widely studied and is currently the major leaching organisms. Leaching is generally done by bacterial action directly on the ore to extract metal or the bacteria produces substances such as ferric iron and sulphuric acid which then extract metal.

8.1 *THIOBACILLUS FERROXIDANS* IN LEACHING

8.1.1 Thiobacillus ferroxidans

Thiobacillus ferroxidans is a gram-negative acidophilic, mesophilic, chemoautotroph. Cells are short, strait rods approximately 1.0 μ long and 0.5 μm in diameter. It tends to live in hot springs, volcanic tissues and sulphide ore deposits. The bacteria gets energy for growth from the oxidation of iron or sulphur. This energy is required for CO_2 fixation and other function. Iron must be in the ferrous or bivalent form (Fe^{++}) and it is converted into ferric form (Fe^{+++}) by the action of bacteria. The soluble and insoluble sulphide forms can be attacked. The elemental sulphur and soluble compounds that incorporate either the thiosulphate ion ($S_2O_3^-$) or the tetrathionate ($S_4O_6^-$). The mechanism of sulfur oxidation by T. ferroxidans in similar to that of other thiobacillus. Energy is produced from oxidation of sulphite to sulphate through enzyme action sulphate oxidase. *Aph* reductase and adenylate kinase.

8.1.1.1 *Mechanism of leaching by T. ferroxidans:* The leaching of metal from ore was long believed to be purely a chemical reaction. But now it is discovered that bacteria is responsible for leaching and it is primarily a biological process. The main leaching mechanism involves indirect leaching, direct leaching and galvanic conversion.

8.1.1.2 *Indirect leaching:* In indirect leaching two important reactions are carried out by *T. ferroxidans*. They are

$$FeS_2 \text{ (pyrite)} + 3.5O_2 + H_2O \rightarrow FeSO_4 + H_2SO_4$$

$$2FeSO_4 + 1.5O_2 + H_2SO_4 \rightarrow Fe_2\ (SO_4)_3 + H_2O.$$

The biologically regenerated ferric sulphate is strong a oxidant capable of dissolving a wide variety of metal sulphide minerals. The sulphuric acid effectively leaches a variety of copper oxide:

$$\underset{\text{(cizurite)}}{Cu_3\ (OH)_2\ (CO_3)_2} + 3H_2SO_4 \rightarrow \underset{\text{(copper sulphate)}}{3CuSO_4} + 2CO_2 + 4H_2O$$

8.1.1.3 *Direct leaching:* Iron oxidising bacterium *T. ferroxidans* may also leach metal sulphide directly without involvement of ferric sulphate. Incubation of wasted cells with these synthetic substrate (sulphides) results in metal solubaliser.

8.1.1.2.4 *Galvanic conversion:* A third leaching mechanism is conversion. This technique involves physical contact between two different metasulphide phases immersed in an electrode (sulphuric and ferric sulphate solution creates a galvanic cell) e.g. in the $CuFeS_2/FeS_2$ complex. Charcoal pyrite has the lower electrical potential and is known as anode and white pyrite behave as a cathode. As a result the chalcopyrite phase undergoes rapid dissolution white pyrite is unaffected. *T. ferroxidans* accelerate the reaction by oxidising the film of elemental sulphur.

8.1.2 *Leptospirillum ferroxidans*

Although *T. ferroxidant* has been the most popular and widely studied as bioleacher, other organisms can also attack mineral sulphide under acidic conditions, one of these are *Leptospirillum ferroxidans*. It was first isolated in 1972 from copper deposit. It is a gram +ve mesophilic, acidophilic, chemolithotropic bacterium capable of ferrous iron oxidation. Recent research showed that *L. ferroxidans* could solubilise pyrite in pure culture and indirectly catalyse leaching of sulphide minerals. Leptospirillium like bacteria have been isolated from mines and mine water at various locations.

8.1.3 *Sulfolobus* sp.

Sulfolobus are the most robust of the leaching microorganisms. They are extremely thermophilic and acidophilic. They can grow autotropically at pH 1-3 and at 50-90°C. *Sulfolobus* oxidises sulphur, ferrous iron and metal sulphide for energy. Leaching studies using *Sulfolobus* has been done in molybdonite, chalconite and chalcopyrite. Sulfolobus solubilise molybdenum and copper from mineral.

Minerals that resist most microorganisms such as chalcopyrite and molybdenite are readily attacked by *Sulfolobus* and solubilise metals. The soluble metals are not toxic to the organisms. Molybdenum is readily soluble to leaching by *Sulfolobus brieylegi* in concentration as high as 75 mg per litre.

How does bacteria leach mineral from rock? To answer this, bacterial leaching has been divided into two approaches, direct and indirect. Direct bacterial leaching involves enzymatic attack by bacteria on components of the mineral that are susceptible to oxidise. In the process of obtaining energy from the inorganic metals, the bacteria triggers electron to be transferred from iron or sulphur to oxygen. The oxidation product is more soluble in indirect leaching, it does not involve any direct attack by enzyme, instead the bacteria generate ferric iron by oxidising soluble, ferrous iron. Ferric iron in turn is a powerful oxidising agent that reacts with other metals and transferring them into the soluble, oxidised form in a sulphuric acid solution. Indirect leaching is usually referred to as bacterially assisted leaching. The control of acidity is important in leaching, an acidic environment must be maintained in order to keep ferric iron and other materials in solvents.

8.2 APPLICATIONS OF MICROBIAL LEACHING

8.2.1 Copper leaching

Copper leaching by biological method is practised in several countries including USA, Chile, Australia, Canada, etc. *T. ferroxidans* is used to leach out copper from the ore.

Copper ore extracted from mines is seggregated. High-grade material is made to concentrate and introduce for smelting, while the lower grade ore is subjected for leaching. The ore is piled on an impermeable surface until a dump is accumulated upto 40 m height. After the top is leached, solution is flooded or sprayed onto the dump. Bacterial colonisation occurs in the top. The temperature may reach 90°C in the centre of the dump owing to exothermic nature of chemical reaction. Leach solution appear at the base of the dump are transported to central recovery facilities. In most large scale operations solution contains 0.5 to 2 gms cm^{-1}, to large units contains iron scapes. The following reaction then occurs as follows:

$$CuSO_4 + Fe^0 \rightarrow FeSO_4 + Cu$$

The finely divided cement copper is regularly recovered and refined for sale.

8.2.2 Uranium Leaching

Uranium leaching using microbes was introduced recently. To initiate leaching process the tunnels of underground mines are flooded with dilute sulphuric acid solution. Bacterial activity is limited to oxidation of pyrite and ferrous iron because *T. ferroxidans* does not directly interact with uranium minerals. Uranium leaching proceeds according to the following reaction

$$UO_2 + Fe_2(SO_4) + 2H_2SO_4 \rightarrow [UO_2\ (SO_4)_3]^4 + 2FeSO_4 + 4H^+$$

the uranium recovering ranges from 30 to 90%. The common yeast *Saccharomyces cereviseae* and fungus *Rhizopus arrhizus* have recently been shown to adsorb uranium from the waste water. The surface binding of uranium is optimised and suggests that positively cleaved uranium ions are attracted to the negatively charged ligands on the cell.

8.3 LEACHING OF OTHER METALS

Copper and uranium are not the only metals that can be recovered by a microbial leaching process. The microbe *T. thiooxidans* has been used in the leaching of cobalt, nickel and zinc. This leaching is indirect. Sulphuric acid is generated during sulphur oxidation by the microbe. Thus leaching of all the metals is an acid solubilise of metals. Microbial leaching of lead results in enrichment of lead in the metal matrix. During oxidation of galena (PbS), the lead is converted to insoluble land surfacts besides, other soluble metal surfacts are also leached, leaving a concentrated lead product. This may be considered as a beneficial process.

8.3.1 Leaching of Precious Metals

Recovery of precious metals like gold and silver from metal ores using microbial system is relatively a recent application. Gold ores are refractory, when the precious metal is encapsulated by iron sulphide minerals. It is unavailable for contact with cyanide. Microorganisms degrade a surrounding matrix, allowing contact between the lixiviant like cyanide and precious metals. Bacterial leaching of the common interfering pyrite and arsenopyrite is well known. Research studies is not centered on the application of this process to specific gold. The comparison was made on gold recovery from cyanide and with bacteria. Interestingly, gold recovered was found to be greater than 90% when pretreated with *T. ferroxidans* and it was only 50% with cyanide treatment.

Microbial pretreatment is an environmentally superior alternative to roasting. The roasting method generates air pollutants such as suiphur dioxide, volatile arsenic compounds and other deleterious compounds or elements. One pilot plant study showed greater gold recovery from microbial pretreatment than roasting. The economic viability of microbial pretreatment will be site specific and will depend primarily on the minerology of the ore.

Table 8.1. Microbial leaching of metals

Metal Sulphides	Leaching microorganisms
CoS	*Thiobacillus ferroxidans* *T. thiooxidans*
ZnS	*T. ferroxidans* *T. thioxidans*
MoS_2	*Sulpholobus brieylegi* etc.
NiS	*T. ferroxidans*

8.4 METAL RECOVERY BY MICROBIAL ACCUMULATION

Either viable species or non viable microbial biomass can be used for recovery of metals from solutions. The living cells can be used to absorb or adsorb metals or by precipitating using metabolic end product. Certain type of microorganisms consist of *Pseudomonas maltophilia, Staphylococcus aureus* and corneform organisms accumulate over 300 mg silver grams per litre (dry wt.). This microbial community was stable and tolerated upto 100 mm silver. A mixed culture of *T. ferroxidans* and *T. thioxidans* can accumulate silver as the mineral acanthite (Ag_2S) during the leaching of sulphide minerals containing silver (Table 8.1).

8.4.1 Removal of Heavy Metals

Effective heavy metal removal was observed in a series of lakes and streams that recovered sewage and effluents from mining. The presence of algae and sulphate reducing bacteria like *Desulfovibrio* sp., *Desulfo tomaculum* sp. and *Desulfomonas pigna* in sewage was responsible for heavy metal removal. Sewage provides inorganic nutrients that stimulates algal growth. This leads to further metal removal through absorption on to biomass as sedimentation of the complex. Production of hydrogen sulphide and reducing conditions were primarily responsible for metal immobilisation in the sediments. This process can be proposed as method for concentrating heavy metal for secondary recovery.

8.5 BIOMINING—A FUTURE STRATEGY

The important programme in bioextraction metallurgy includes

(i) Effective microbial desulphurization of coal.
(ii) Microbial removal of methane from coal mines.
(iii) Immobilisation of pollutants of toxic elements such as arsenic or mercury.
(iv) Microbial process and removal of pyrite impurity from pottery clay.
(v) Methane from organic residue waste.

Based on laboratory studies done by many researchers on current applications in the field, it is clear that microorganisms and the diversified activities will help to lay claim to mineral wealth buried deep in the ground. These small servants of man promise to help in changing the air and water, leaching valuable metal resource.

QUESTIONS

1. Discuss the use of microorganisms in recovery of metals from mines.
2. What is Bioleaching? Discuss the mechanism of bioleaching of ores by microbial action.

SHORT QUESTIONS

1. Bioleaching (Nov. 1997) (Nov. 1999)
2. Thiobacillus ferroxidans
3. Metal reclamation
4. Biohydrometallurgy

9

Biofertilisers

India has contrived second green revolution long ago. Despite the progress the country's need of food production is increasing to a larger extent. Contribution of fertilisers in increasing crop productivity reflects country's self sufficiency in food production. Use of chemical fertilisers have been in practice since many decades and this has resulted in a manifold increase in crop yield. But the adverse effect of chemical fertilisers in the long run have been noticed on soil, quality of water, food, etc. The indiscriminate use of inorganic chemicals are already depleting useful soil micro-flora. Continuous use of high doses of fertilisers along with pesticide in agriculture could be detrimental and cause a major environmental pollution. Hence, organic fertilisers or biofertilisers are very attractive, safe and ecofriendly. It can be a good substitute for chemical fertiliser to sustain agriculture, where it involves effective manage-ment of agri resources for the benefit of mankind while maintaining environmental quality. Sustainable agriculture depends largely on renewable resource which is achieved by biological nitrogen fixation.

According to one estimation at least 250 million tonnes of food grains will be required to meet the challenge posed by 1 billion population. Currently 15 million tonnes of fertilisers are used in agriculture field. The present use of 13 million tonnes of fertilisers per annum would be doubled to meet the target of 250 million tonnes of food production. Hence, integrated use of all fertilisers such as chemical fertilisers, organic fertilisers and biofertilisers are indispensable as an alternative for improving soil fertility. Biofertilisers in combination with green manure can efficiently combat the ill effect of chemical fertilisers. Biofertilisers are the product containing living cells of different types of microorganisms which are able to provide non usable to usable nutritional elements by biological process. Otherwise biofertilisers are simply microbial inoculants which are able to fix atmospheric nitrogen.

9.1 CLASSIFICATION OF BIOFERTILISERS

Biofertilisers can be broadly classified into (i) Nitrogen fixing biofertilisers (ii) phosphate mobilising biofertilisers (iii) Cellulose or organic matter decomposition biofertiliser.

9.1.1 Nitrogen fixing Biofertiliser (Bacterial biofertiliser)

Nitrogen occupies 78% of total atmospheric gas. Plants are unable to utilise nitrogen in gaseous form. Thereby nitrogen is essentially reduced to nitrate and ammonia by a biological process. This nitrogen conversion is generally known as biological nitrogen fixation. Several microorganisms in nature are responsible for converting unusable nitrogen to usable nitrogen by plants. The organisms responsible for nitrogen fixation are known as nitrogen fixers or Diazotrophs. Some of the common nitrogen fixers which inhabit, soil are *Rhizobium, Azotobacter, Clostridium* and blue green algae.

9.1.2 Biological Nitrogen Fixation (Symbiotic)

Biological nitrogen fixation involves the reduction of nitrogen to ammonia. This becomes incorporated into the organic nitrogen by nitrogen fixing organisms. The oxidation of NH_4^+ to nitrite is mediated by the bacterium *Nitrosomonas* and nitrite to nitrate by *Nitrobacter*. The nitrate which is absorbed by the roots of plants must first be reduced to ammonium before it is incorporated into plants.

The symbiotic nitrogen fixation is carried out by two organisms which are living together and mutually beneficial. The best example is legume *Rhizobium* and *Anabaena-Azolla* symbiosis.

9.1.2.1 *Rhizobium as Biofertilisers:* The genus *Rhizobium* belongs to the family Rhizobiaceae and divided into three main groups: Azorhizobium, produces stem nodulation in plants. Second group is *Bradyrhizobium* which contains two species, *B. japonicum* and *Bradyrhizobium* species produce root nodules in leguminous plants. Rhizobia are gram –ve, rod shaped, non spore forming motile bacteria. They use monosaccharides and disaccharides as carbon source.

Rhizobium is a very useful biofertiliser. The feasibility of this bacteria in fixing nitrogen to the soil was realised long ago. They interact with specific leguminous plants and produce nodulation in the roots. Here the nodule acts as a mini factories for ammonium production, which are assimilated nitrogen readily available to the plant cells. Scientists developed microbial inoculants as a suitable biofertiliser (Table 9.1).

Rhizobial inoculants can be prepared from the Rhizobia of the root nodules after isolating, culturing and mass-production. It was then inoculated into peat powder and applied to the soil. The Rhizobium – legume association can fix upto 100-250 kgN/ha in one crop season. However, inoculation of a particular strain of Rhizobium to a particular variety of legume increases crop yield. In order to obtain 1000 kg of *Rhizobium* inoculant, approximately 15,000 tonnes of Rhizobium culture are required.

Table 9.1. Host for nitrogen fixing microorganism.

Host	Nitrogen fixing bacterial species
1. Peas	*Rhizobium leguminosarum*
2. Clover	*Rhizobium trifoli*
3. Mung beans	*Rhizobium phaseoli*
4. Soya bean	*Bradyrhizobium japonicum*
5. alfa alfa	*Rhizobium meliloti*

9.1.2.2 *Rhizobium inoculation production:* The *Rhizobium* is grown in 3 litres of yeast mannitol broth in 5 litres conical flask on a rotary shaker. When the primary culture is ready, it is then poured into the fermentor for mass multiplication during 3 or 4 days. Batch culture is more suitable than continuous culture for the production of *Rhizobium* inoculum. Rhizobium in the fermenter is emptied through a media dispensor. Meanwhile peat is taken in polypropylene bags and are then autoclaved for 60 minutes two consecutive days. The peat bag is wiped with alcohol and 80 ml *Rhizobium* inoculation is injected quickly into 120 g peat bag and the injected hole is sealed immediately. These bags are kept at room temperature for incubation for at least a week.

9.1.3 Asymbiotic Biological Nitrogen Fixation

Biological nitrogen fixation is done by organisms which live freely in soil without any symbiotic association. Some of the organisms which are intensively used as nonsymbiotic fixers are *Azotobacter, Azospirillum*, and among algae *Anabaena* of blue green algae is used in paddy fields.

9.1.3.1 *Azotobacter:* There are six species of *Azotobacter, A. chrococcum* is the most commonly used species. The bacterium is gram negative, ovoid in shape accumulate poly, β-hydroxybutyrate granules. They use mono, di and polysaccharides as carbon source.

Azotobacter is aerobic free living, fix nitrogen in non symbiotic fashion. It requires pH ranges of neutral to alkaline. They excrete ammonia in the rhizosphere in presence of root exudates. They can even produce plant growth promoting hormones especially Indole acetic acid.

Azotobacter Inoculum production: *Azotobacter* sp. can be isolated from soil by serial dilution method. These are then cultured in Azotobacter specific Ashbylsensins nitrogen free agar media and poured into sterile petriplates. The plates are incubated at 30°C for 3-4 days. These organisms appear as dew drops on milky agar plates (Table 9.2). The plates are incubated for

30°C for 3-4 days. Once primary culture is over, it is then transferred to batch fermentor for large scale production.

Table 9.2. Nutrient media for nitrogen fixing organisms

	Media component	Yeast mannitol agar media (Rhizobium) g/lt	Ashby's mannitol agar media (Azotobacter) g/lt	Prigshem's media (Blue green algae)
1.	Mannitol	10.0	20.0	–
2.	K_2HPO_4	0.5	0.2	–
3.	KNO_3	–	0.1	0.02 %
4.	$MgSO_4.7H_2O$	0.2	0.2	0.001 %
5.	NaCl	0.1	0.2	0.002 %
6.	$(NH_4)_2$ HPO_4	–	–	0.002 %
7.	$CaCO_3$	–	5.0	–
8.	$CaCl_2.6H_2O$	–	–	0.0005 %
9.	$FeCl_2$	–	–	0.00005 %
10.	Yeast Extract	1.0	–	–
11.	Agar	20	15	–

9.1.3.2 *Azospirillum:* It is a gram positive motile bacterium which produces poly-β-hydroxybutyrate. The low oxygen tension is essential for nitrogen fixation by *Azospirillum*. Azotobacter induce plants to excrete mucilage to reduce oxygen tension in the rhizosphere and helps to fix atmospheric nitrogen.

9.1.4 Mechanism of nitrogen fixation

Conversion of atmospheric nitrogen into ammonia requires high amount of energy (N ≡ N). (In industries conversion is done by high temperature and pressure). Biological system has a unique potential in converting atmospheric nitrogen into nitrate and reduced to ammonia in an ordinary cellular temperature using ATP energy. All nitrogen fixing microorganisms including blue green algae contain a special group of genes known as nif genes which helps in cellular conversion of nitrogen.

9.1.4.1 *Root Nodulation:* The root nodulation in legumes begins when some of the chemical sustances are released during seed germination. These chemicals would attract *Rhizobium* and begins to develop infection threads. The enzyme of bacteria begins the infection threads. Bacterial enzyme damages the part of the cell wall and allows it to enter into root hair cell which causes root hair to produce infection thread. The first symptom of infection is branches of moderate root hair curling. All rhizobia can infect root hair cell. The infection thread is the site for extensive multiplication of *Rhizobium*. In the inner cortex cell, the bacteria are released in the cytoplasm and stimulate other cells to divide. This results in the rapid proliferation of the tissue and it is transformed into root nodule. Root nodules contain several thousands of bacteroids. Nitrogen fixation in root nodules occur exclusively within the bacteroids. Some of the electrons and ATP obtained during oxidation by the bacteroids are used to reduce nitrogen to ammonia. The enzyme nitrogenase catalyses this reaction.

9.2.4.2 *Nitrogenase:* All nitrogen fixing (Symbiont and Asymbionts) organisms and blue green algae possess a unique set of clustered genes known as nif genes which produce an enzyme, nitrogenase. This enzyme in bacteria fix nitrogen by converting nitrogen into ammonia using hydrogen ion and energy by host plant. Reduction of nitrogen to ammonia requires electrons which are then transferred to nitrogen by flavodoxin and ferrodoxin transport proteins. Twelve to fifteen ATP molecules are consumed per nitrogen molecule.

$$N_2 + 8H^+ + 8e^- + 16 \text{ mg ATP} \rightarrow 2NH_3 + H_2 \uparrow + 16 \text{ mg ADP} + 16 \text{ P.}$$

The nitrogenase enzyme consists of two complexes I and II. The complex one consists of two α protein subunits of approximately 50,000 daltons. It also associates with β-subunit of 60,000 daltons, 24 iron and 2 molybdenum molecules. The complex II comprises of 22 subunits of 32,000 daltons. Component I catalyses reduction of N_2 and component II donates electron to component I.

The genome of nitrogen fixing organisms consists of unique set of nearly 20 nif genes organised into seven operons. The nif genes of *Klebsiella pnuemoniae* has been sequenced completely.

QB	Al	F	MZWVSV	XNE	YTKDH	J

nif gene operon

The genes nif H, nif D and nif K encodes three structural polypeptide of nitrogenase. Nif B, N, E and V are involved in the synthesis of Femo cofactor and Nif F and Nif J provide a flavodoxin and a pyruvate.

9.3 BLUE GREEN ALGAE AS A BIOFERTILISER

Blue green algae (BGA) is otherwise known as cyanobacteria (Table 9.3). These are free photosynthetic nitrogen fixers. Several investigations confirm the role of BGA in enhancing soil fertility. Almost all BGA are distributed worldwide. Blue green algae are unicellular, branched or unbranched filaments algae and contain a specialised structure called heterocyst. This is the site for nitrogen fixation.

BGA is used prominently in paddy fields to increase soil fertility. The study conducted by Central Rice Research Institute, Cuttack, emphasised that BGA contributes to the overall grain yield of rice from 6 to 75%.

Table 9.3. Types of BGA

Anabaena	*Scytonema*
Nostoc	*Aulosira*
Plectonema	*Tolypothrix*

9.4 PHOSPHATE SOLUBULISING MICROORGANISMS

Phosphate is the only next key element after nitrogen in plant growth development. About 90% of Indian soil is poor in the available phosphorous. This lacunae is fulfilled by adding soluble phosphate, which around 10-15% is utilised by crop plants. Phosphorus is found in soil, manure and plant in various organic and inorganic forms but are unavailable to the plants. Phosphate solubility is a serious problem in agriculture practice.

Various microorganisms have reported to solubilise insoluble inorganic phosphates in soil environment and facilitate the uptake process. The list of various phosphate solubilising microorganisms given in Table 9.4. Several factors responsible for solubilisation process are energy, pH and temperature. The pH range 4-5 is ideal for certain phosphate solubilising fungi and the optimum temperature for fungus in 30°C whereas for bacteria is 20-30°C.

Phosphate solubilising microorganisms excretes certain organic acids like citric, glutamic, lactic, oxalic and fumaric acid etc. These organisms help in weathering of rocks by acidity and reduce the particle size to almost amorphous form. The action of organic acids has been linked to their ability to form stable complex with metallic elements. They can even mineralise organic phosphorus into soluble form, all the reactions are confined to rhizosphere. There are more than one chemical process involved in phosphate solubilisation. In addition to chelation of metals, organic acids have acidifying effect. *Aspergillus niger* was found to bring ferric sulphate into solution by production of H_2S gas. It reduces insoluble ferric phosphate to ferrous sulphide. This soluble form is readily available upto usable form by plant.

Table 9.4. List of phosphate solubilising organisms

Bacteria	Fungi	Yeast
Bacillus cereus	*Aspergillus awamori*	*Torula thermophila*
B. polymixa	*A. flavus*	*Sehwannio*
Aerobacter	*A. niger*	*Myces occidentalis*
Agrobacterium	*Pencillium* sp.	
Azotobacter	*Rhizoctonia*	
Chlorocacum	*Rhizopus*	
B. subtilis	*Sclerotium rolfsic*	
Erwinia sp.		
Pseudomonas sp.		

9.5 VESICULAR ARBUSCULAR MYCORRHIZAE (VAM)

The symbiotic association between fungal mycelia and plant root is known as mycorrhizae. The best known fungus for symbiotic association are vesicular arbuscular mycorrhizae (VAM). This mycorrhizae involves significantly in Phosphorus cycle and uptake of 'P'. VAM acts as an efficient biofertiliser. It colonises the root and increases the growth of the crop. Apart from involving 'P' uptake. VAM can also increase nutrient uptake like Zn, Cu and helps in the synthesis of plant growth promoter, tolerant to drought, salinity, disease resistance etc.

VAM fungi are worldwide in their distribution from aquatic to desert environment. These fungi can be cultured on media as they are obligate symbionts (grow only with living cells). Most important species of VAM are Glomus, Sclerocystis, Acaulospora, Endogene etc.

9.5.1 VAM Inoculum production

The soil culture was the most preferable method for isolation of mychorrhizae. The soil or root from the field are inoculated into roots of plants growing in sterilised soil. This is known as pot culture. To obtain starter culture, spores are removed from the fielded pot culture by density gradient method or using wet sieving. Healthy spores are collected and sterilised by using 2% sodium hypochlorite, 200 ppm streptomycis followed by rinsing in sterile water. About 10-20 spores are inoculated on the seedlings growing under green house condition to facilitate the colonisation of the mycorrhizae. To harvest VAM inoculants, the roots are taken after prooning plant tops and then chopped into uniform inoculants before storage. Generally maize, sorghum, onion etc. are being employed for VAM innoculum production. VAM inoculum can also be produced in peat, saw dust and vermiculate.

9.5.2 Significance of VAM as Biofertiliser

VAM has good potential in importing nutrient cycles. Significant yield increase with VAM were reported in groundnut and chillies. In other studies, considerable crop yield increase was noticed over control by inoculating VAM in soyabean. Many people are of the opinion that inoculation of VAM plus Rhizobium was more advantageous than using single inoculum. VAM fungi can save fertiliser upto 30% besides improving crop yield.

QUESTIONS

1. What are Biofertilisers? Explain their role in agriculture.
2. Classify biofertilisers. Describe their importance in crop production.
3. Describe mechanism of nitrogen fixation. Add a note on its significance.

   BU (April 2000)

SHORT QUESTIONS

1. Bacterial biofertilisers (BU April 2000)
2. Biofertilisers (BU Nov. 1999) (Nov. 1997)
3. *Rhizobium* (May 1996)
4. *Azolla* (BU Nov. 1999)
5. VAM, fungi
6. Blue Green Algae
7. Symbiotic nitrogen fixation
8. Non-symbionts

10

Environmental Significance of Transgenic Plants & Animals

Breeders in the past have been restricted in the use of genetic resources and concentrating on variety improvement. The genetic resources present in other plants have now become accessible for breeding because of advances in recombinant DNA technology, the science of studying and manipulating genetic material. It is now possible to produce transgenic organisms containing gene(s) from novel source. Improvement of fibres and wool from living organisms (plants and animals) is a challenging task. Fibre and wool are the most economically important products from plants and animals respectively. Several considerations have shown through conventional breeding technique, but their quantum of success is still far reaching. On the contrary gene manipulation technology is straight forward and could be targeted to one particular character like fibre and wool. The techniques have recently been extended to cotton and sheep, opening up tremendous possibilities for the improvement of cotton fibre and wool from the sheep.

10.1 INSECT TOLERANT TRANSGENIC COTTON

The *Agrobacterium* provides a novel system for transferring novel genes into cotton plant, either by using binary vector or cointegrative method. Novel genes can be cloned into chromosomal DNA of plant. More description was emphasised in recombinant DNA technology (Part A).

Once novel gene is introduced into cotton through tissue culture system, miniature transgenic plants can be produced. Cotton has proven to be difficult species because it responds poorly in tissue culture, although several laboratories have had some good success. Today useful genes for cotton improvement is mainly protecting cotton from insects, diseases and herbicides. Cottons are attacked by a number of insect pests, but the major ones are the caterpillars of the cotton budworm (*Helicoverpa armigera* and native bud worm *H. punctigera*). The non environmentally friendly synthetic pyrethrin insecticides have been effective in the past but there are growing fears that resistance development by insects may soon make pyrethroids ineffective. This, together with public concern on environment has led to a development of more insect tolerant cotton variety.

The *Agrobacterium* provides a way of getting novel genes into a single cell of a plant. The tissue culture system that allows the proliferation of a single cell and the regeneration of whole plant.

10.2 HERBICIDE TOLERANT PLANTS

Genetic engineering of plants for crop improvement was first demonstrated by developing plants with resistance to commercial herbicides. A good herbicide can selectively kill weeds without affecting crop plants and should not be toxic to other organisms. But none of the commercial herbicide fulfils the above criteria. Hence transgenic plant with a herbicide tolerance is a novel approach. In addition, accumulation of herbicide in the soil can kill useful micro-organisms. To circumvent this problem novel approach have been made to produce herbicide resistant plant in which, gene that codes for an enzyme that modifies herbicides and in inactive.

Glyphosate is a herbicide that kills the plants rapidly when applied at low concentrations. Glyphosate can inhibit EPSP (5-enolpyruvyl shikimic acid-3-phosphate synthase) enzyme which catalyse the synthesis of several aromatic amino acids through Shikimic acid pathway. EPSP synthase in plants is coded by a nuclear gene. An EPSP synthase gene which conferred resistance to herbicide glyphosate was first found in a mutant of bacterium *Salmonella* that had a mutation in aro A locus that encodes EPSP synthase. In genetic engineering strategy A chimeric gene was made with a octopine synthase promoter, mutant aro A coding sequence (aro A isolated from *Salmonella* bacterium) and poly A signal. This gene construct was introduced into tobacco plants using *Agrobacterium* Ti plasmid. Transgenic tobacco plants expressed high level of bacterial EPSP synthase resistance to herbicide glyphosate.

2,4-Dichlorophenoxy acetic acid is a cheap weedicide, and the cotton plant is extremely sensitive to this chemical and causes extensive damage to the crop even at distances over 60 km from the initial site of application. Genetic engineering offers the potential for a solution to this problem of 2,4-D spray damage to cotton, by making the cotton plants biochemically tolerant to 2,4-D. The main genetic engineering strategy is to introduce a gene taken from the bacterium *Alcaligenes entrophus* into cotton plant for 2,4-D resistance. *Alcaligenes* contains a number of genes essential for the complete breakdown of 2,4-D. First step in the complex pathway of 2,4-D degradation may be sufficient to detoxify 2,4-D in plants.

10.3 ENVIRONMENTAL IMPACT OF TRANSGENIC PLANTS

One concern about the risk of transgenic cotton is the escape of transgenes through pollen dispersal (via wind, insect) from transgenic crop plants to their relatives. Dispersal of pollen from transgenic cotton plants to surrounding non-transgenic plants have been observed.

10.3.1 Out-cross of herbicide-tolerant plant

The escape of transgenes through pollen is regarded as an environmental concern in the case of herbicide resistance genes resulting in "Superweeds", which are resistant even to broad spectrum herbicides. According to recent study cotton is likely to hybridize with wild congeners. Recently, the resistance of a weed to three herbicides has been reported which derived from transgenes in canola.

10.3.2 Environmental Impact of Terminator gene technology

Delta and Pine Co., USA and the US Department of Agriculture received patent on a new genetic technology designed to prevent unauthorized seed saving by farmers. The technology is "control of plant gene expression". In other words, the new technology genetically alters the seed so that it will not germinate if replanted for a second time. This new technology is known as "Terminator gene Technology".

10.3.2.1 *Terminator—How it works:* The key to the terminator technology is the ability to make a toxin confined to seeds and kill the embryo in last stages of development. To accomplish this, the plan is to take the promoter from a gene normally activated late in seed development and to fuse that promoter to the coding sequence of a protein that will kill an embryo going through the last stages of development.

To accomplish the terminator effect in plant, three engineered components be transferred into the plant DNA.

(a) a toxin or lethal gene controlled by a seed specific promoter but blocked by a piece of DNA in between the promoter and coding sequence.

(b) a repressor protein coding sequence with a promoter that is active all of the time.

(c) a recombinase coding sequence, controlled by a promoter that would be active at all times, except that it is regulated by repressor protein which can be regulated by tetracycline.

10.3.2.2 *Will the Terminator spread to other plants*? It is likely that terminator will kill the seeds of neighbouring plants of the sane species under certain conditions. However, the effect will be confined to the first generation. When the seeds germinate into plants and produce pollen, every pollen grain will carry a ready-to act toxin gene. If the terminator crop is next to a field planted in a normal variety and pollen is taken by insects or the wind to that field, any eggs fertilized by the terminator pollen will now have one toxin gene. It will be activated late in that seed's development and the seeds will die. Besides, will the seeds containing the toxin made by terminator be safe to eat is a million dollar question. The toxin could end up in a product without any one's knowledge. For example, an ornamental sunflower could spread terminator to an oilseed variety, and then to toxin could end up in edible oil. Other potential problem is edible seeds containing this toxic protein have to do with allergy.

10.4 TRANSGENIC SHEEP

Australia and Newzealand are the largest producers of wool and is an important source of income. Wool is of immense value in the manufacture of fabrics. The pure wool in the fabrics exhibit thermal insulation.

Wool is basically a keratin protein and contains specific amino acids like cysteine. The gene for these aminoacids have been cloned in sheep embryo to produce transgenic sheep. The two enzymes serene acetyl transferase (SAT) and O-acetyl serine sulphydrylase (GAS) is involved in cysteine biosynthesis. The genes for this enzyme CySE and Cysm were isolated from bacteria and inserted into a vector. In another approach two genes ace A and ace B from *E. coli* are introduced into animals product of the two genes expression are involved in glucose accumulation. This would provide sufficient fuel molecule needed for wool growth.

10.4.1 Sheep contains genetically altered rumen bacteria

Certain bacteria like *Butyrivibrio, Ruminococcus* which inhabits rumen of the sheep, helps in the breakdown of cellulose. The enzyme cellulase synthesised by these bacteria attack plant fibres, degrade cellulose, hemicellulose to sugars. Simultaneously plant proteins are hydrolysed into amino acids. These amino acid pool can be cycled back to resynthesis of bacterial proteins. The sheep can also use amino acid reserve for its keratin protein.

10.4.1.2 *Transgenic plants favours wool production:* For better growth and production of wool, sheep require sufficient reserve of particular amino acids. Sheep feed on plants; if the plants are genetically engineered to produce proteins with high amount of specific amino acid, it may facilitate increase wool production in sheep. Certain seed specific proteins like pea albumin and vicilin rich in cysteine amino acids, which are required for keratin wool protein, have been isolated and successfully cloned in tobacco plant as a model system. These two genes are expressed in tobacco leaves. Several attempts have been made to extend this technique to other fodder plants so that they would be able to provide sufficient amino acid pool for wool production.

Transgenic sheep is an ideal choice for animal farming. Milk from transgenic sheep is an excellent source from which proteins can be purified.

Several projects have been launched for the production of transgenic sheep worldwide. The UK based PPL company determined to increase the number of transgenic sheep from the present 50 to 1000 by the year 2002. One of the company's leading product is α-antitrypsin (AAT). AAT can be easily purified in large quantities from the milk of transgenic sheep. This product is scheduled for licencing for commercial product. PPL company in Newzealand "Poll Dorset Sheep which gives upto 400 litres of milk per year on twice a daily milking, express foreign proteins at a level of up to 40g per litre. Annual quantities of 1 to 100 kg of proteins can thus be easily obtained.

QUESTIONS

1. Discuss the significance of "Transgenic plants". (BU, April 2001)
2. Enumerate the environmental significance of transgenic organisms.

SHORT QUESTIONS

1. Transgenic sheep (BU, April 2001)
2. Transgenic cotton
3. Terminator gene technology

Part C
Animal Cell Culture

1

History and Scope of Animal Cell and Tissue Culture

Cell culture is essentially a novel and exciting technology in the field of modern biology. Currently, this technology is indispensable in several branches of life sciences. Cell culture techniques has laid a foundation for studying the regulation of cell proliferation, differentiation and product formation in carefully controlled conditions. Cell culture has also benefitted in studying human genome by understanding the pathways of intracellular and intercellular signalling which ultimately regulate gene expression. It acts as indispensable tool for scientist from different fields like immunology, bioengineering, molecular genetics and medicine.

Harrison (1907) and Carrel (1912) for the first time inaugurated and devised tissue culture technique in order to study the behaviour of animal cells. In 1907, Harrison preferred to select frog as a source of tissue for his studies. The preference for frog, because it was a cold blooded animal and further, incubation was not required. In addition, tissue regeneration was more predominant in lower vertebrates than Mammalian tissue. Hence, he thought of applying to mammalian tissue.

Although the cultivation of tissue *in vitro* by Harrison may have created new wave of interest. The medical science extended further interest into warm blooded animals, where normal and pathological development are closer to human. The most favourite choice for tissue culture and embryonated hen's egg. Further, rapid progress in the development of experimental animal husbandry, with genetically pure strains of rodents has made mammals as a comprehensive favourite material for tissue culture. Eark et al 1943, considered rodents tissue had the advantage of producing continuous cell lines and considerable stock transplantable tumors. According to Gey et al 1952, human tumors could also contribute in producing continuous cell line. The classical example is Hela, this encouraged interest in human tissue, Puck and

Marcus (1955) had established method for cloning cells by dilution technique. In 1959 Lovelock and Bishop demonstrated the efficacy of DMSO (Dimethyl Sulfoxide) in cryopreservation of cells.

The development of tissue culture requires two major branches of medical science. One is the production of antiviral vaccines and another to extensive knowledge on neoplasia. the standardisation of condition by cell line. The production of assay of viruses provided immense value and impetus to the rapid development of modern tissue culture technology in the production of large number of cells for biochemical analysis.

This and other technological leap made possible by the availability of media and sera. Besides, control contamination by antibiotics and successful establishment of aseptic culture has made tissue culture an accessible to a wide range of interest.

The exploitation of animals for cell and tissue culture studies triggered controversies and objections from animal right activist. This led to severe handicap in conducting experiments. But animals are very much essential for preclinical trials of new pharmaceuticals. However, use of animals for development of cosmetics and for other activities could not have been justified. Hence, there is growing demand for more extensive *in vitro* assays, the adaptation of which may requires their proper validation. The unprecedented development of very sensitive and more readily performed *in vitro* assays, together with real prospect of assaying for inflammation *in vitro* has promoted wide expansion in *in vitro* testing.

In addition to virology and cancer research several other areas of research have come to depend heavily on tissue culture techniques. The introduction of cell fusion technique by Barski et al in 1961 and genetic manipulation by Frederick et al in 1993 has established somatic cell genetics as a major area in the genetic analysis of higher animals. In the same line of work the most classical example is the contribution of monoclonal antibodies technique in the field of immunology and production of hemopoetic cell lines. The introduction of novel genes into normal human cells to produce pharmaceutically viable drugs. In addition, cell products such as insulin, interferon, and human growth hormones have been genetically engineered. Other areas of major interest include the study of cell interactions and intracellular control mechanism in cell differentiation and development, besides analysis of nervous functions. The progress in neurological research was not in expected line however, of working with propagated cell lines, as propagation of neurons has not been possible so far *in vitro* methods without resorting to the use of transformed cells.

Efficacy of tissue culture technology has been widely used in routine applications in medicine and industry. The chromosomal analysis of cells derived from womb reveals genetic disorders in the unborn child. Viral infections may be assayed on monolayer of appropriate host cells. In addition, several other parameters includes toxic effect of pharmaceutical compounds and environmental pollution can be measured using *in vitro* assays. Further applications of cell and tissue culture includes homografting and reconstructive surgery using an individual's own cells and it has now become accepted clinical practice in some burn units to biopsy and graft the cultured cells back into the areas of most severe burns.

Defective cells can be rectified by transfecting normal genes into genetically deficient cells. Thus, it has become possible to graft such rectified cells back into the patients. The scope for implementation of normal cells from adult or fetal tissue matched donors or genetically reconstituted cell from the same patient is now a reality.

1.1 MERITS AND DEMERITS OF TISSUE CULTURE

Although tissue culture offers multiple advantages. There are several demerits are associated. Hence, extreme cautions to be taken to avoid indiscriminate use of tissue culture. The advantages and disadvantages of cell and tissue culture are listed in the Table 1.1.

Table 1.1. Advantages and Disadvantages of the use of animal cell and tissue culture

Advantages	Disadvantages
1. Controlled physiochemical environment. (pH, temperature, osmotic pressure, O_2, CO_2, etc.)	1. Expertise is needed, so that behaviour of cells in culture can be interpreted and regulated.
2. Controlled and defined physiological conditions (constitution of medium, etc.)	2. Ten times more expensive for some quantity of animal tissue, therefore reasons for its use should be compelling.
3. Homogeneity of cell types (achieved through serial passages)	3. Unstable aneuploid chromosome constitution.
4. Economical, since smaller quantities of reagents are needed than *in vitro*.	
5. Legal, moral and ethical questions of animal experimentation are avoided.	

QUESTIONS

1. Discuss the history of animal cell culture.
2. Enumerate the scope of animal cell culture.
3. Discuss the merits and demerits of animal cell culture.

SHORT QUESTIONS

1. History of animal cell culture. (BU April 2001)

2

Laboratory Facilities

The laboratory facilities for animal cell culture requires certain specific needs. Based on the requirements it can be classified into (a) essential (b) beneficial and (c) useful (Table 2.1).

2.1 ESSENTIAL EQUIPMENT

(i) *Incubator:* the incubator consist of forced air circulation, provision for temperature control of ± 0.5°C and a safety thermostat. Knob, that regulates the temperature to avoid over heating of the incubator. It should be made up of a corrosion resistant material which could be easily cleaned. The incubator consist of double cabinet one above the other, preferably one large cabinet which are independently regulated. Many incubators are provided with a hot water Jacket for even distribution of heat around the cabinet, which holds the temperature, if the heater fails or if there is power failure.

(ii) *Sterilizer:* A domestic pressure cooker is the simplest and cheapest sterilizer. The autoclaves, which can accommodate more capacity can be considered. The simple bench top autoclaves will be sufficient, but a larger model with a timer given more capacity in use.

(iii) *Refrigerators or Freezers:* The domestic refrigerators without the ice box can be used, defrosting is not required which gives more space. If more accommodation is required then the Blood Bank or catering refrigerators are preferred. If there is sufficient space, more than four people are working on tissue culture then, installation of cold room is recommended. Cold room will be more economical in space than several separate refrigerators and is easily accessible. The walls and shelves of the cold room should be smooth and should be washed regularly to minimize the fungal contamination. Most of the tissue culture reagents can be stored at –20°C.

For the preservation of enzymes and antibiotics auto defrost freezers are not recommended. This may cause severe cryogenic damage to them especially

Table 2.1. Tissue Culture Equipments (Freshney 1987)

Minimum Requirement (essential)	Desirable Features (beneficial)	Useful Additions
1. Incubator	1. Laminar flow hood (s)	1. –70°C freezer
2. Sterilizer (autoclave, pressure cooker, oven)	2. Cell counter	2. Glassware washing machine
3. Refrigerator	3. Vacuum pump	3. Closed-circuit TV for inverted microscope (s)
4. Freezer (for –20°C storage)	4. CO_2 incubator	4. Colony counter
5. Inverted microscope	5. Coarse and fine balance	5. High capacity centrifuge (6 × 1 litre)
6. Soaking bath or sink	6. pH meter	6. Cell sizer (eg. Coulter ZB series)
7. Deep washing sink	7. Osmometer	7. Time-lapse cine-micrographic equipment
8. Pipette cylinder (s)	8. Phase-contrast and fluorescence microscope (s)	8. Interference-contrast microscope
9. Pipette washer	9. Ortable temperature recorder	9. Polythene bag scaler (for packaging sterile items for of log-term storage)
10. Water purifier	10. Permanent temperature and recorders on sterilizing oven autoclave	10. Controlled rate cooler (for cell freezing)
11. Bench centrifuge	11. Roller racks for roller bottle culture	11. Filing for freezer records and catalogues
12. Liquid N_2 freezer (~ 351, 1,500-3,000 ampoules)	12. Magnetic stirrer racks for suspension cultures	12. Centrifugal elutriator centrifuge and rotor
13. Liquid N_2 storage flask (~ 251)	13. Pipette drier	13. Fluorescence-activated cell sorter
	14. Pipette plugger	14. Densitometer
	15. Trolleys for collecting soiled glassware and redistributing fresh supplier	15. Density meter (for density gradient cell separation)
	16. Pipette aid	
	17. Autopipette or other form of automatic dispenser diluter	
	18. Separate sterilizing oven and drying oven	

serum will deteriorate due to oscillation in the temperature of an auto defrost freezer.

(iv) *Inverted Microscope:* A simple inverted microscope is used to analyse morphological changes in the cultured cells and also to observe the characteristic pattern of microbiological infection. The microscope stage should enough space to observe big roller bottles. For better photography of living cultures, the microscope should have high quality optics, a long working distance phase contrast condenser and objectives with provision to take an image. The classical example is olymphus CM. The nikon marking ring is a useful accessory to the inverted microscope. This devise is placed on the objectives to locate the desired colony or a patch of cells by marking the underside of the dish.

2.2 WASHING FACILITIES

(i) *Sinks for washing purpose:* The stainless steel or polypropylene sinks of required sizes can be built. All the glasswares can be immersed in detergents during cleaning or washing. The stainless steel is used for radioisotopes and polypropylene is used for hypochlorite disinfectants. Washing sinks are provided with taps to get cold water and hot and cold mixture. Besides stainless steel tap for deionised water from reservoir, above the sink.

(ii) *Pipette cylinders:* Pipetti cylinders are made from polypropylene to stand for 2 hrs in disinfectant, before washing.

(iii) *Pipette washer:* Pipette should be soaked in detergent first before finally rinsing in deionized water.

(iv) *Pipette drier:* Pipettes can be dried in regular oven or using the electric drier.

(v) *Sterilising and drying:* Though the autoclave is used for most sterilization purposes it is preferable to sterilize pipettes and other glasswares by dry heat to avoid possibility of chemical contamination. Dry heat sterilization is done at high temperature of 160-180°C. Two small ovens can be used instead of bigger one, as heating is easier, more uniform and more economical when only little glasswares are being used.

(vi) *Water purification:* Deionized water is required for rinsing glasswares. Media preparation requires higher degree of purity. Water can be purified by three or four stage processes. These are filtration, deionization, millipore filtration and distillation. Distillation process can produce sterile water.

Purified water can not be stored, but recycled through apparatus continuously to minimize algal contamination. Water is the critical constituents of media and reagents particularly for serum free media.

2.3 BENEFICIAL EQUIPMENTS

The above discussed equipments are essential equipments for a modern tissue culture laboratory. But there are several items of equipments that will make the laboratory efficient and more easier to use and also better controlled. If the facility is to be used frequently by more than two people then, many of the following equipments will add to the essential list.

(i) *Laminar Flow:* The introduction of laminar flow cabinets or woods to provide a clean bench has greatly facilitated the maintenance of aseptic and sterile conditions, so that no separate room for sterile handling is needed and a laminar flow cabinet can be kept in an unspecialized laboratory space. Usually one wood is sufficient for two to three people and horizontal flow hood is cheaper and gives best sterile protection to the cultures. To work with hazardous materials like radioisotopes, carcinogenic or toxic drugs. Virus producing cultures or any primate cell lines a class II biohazard cabinet should be used.

Size of 4ft × 2ft hood is preferably considered for one person to use at a time. It should be easily cleaned both inside the working area and below the work surface in the event of spillage. The front screen of laminar flow should be able to be raised, lowered or removal completely to facilitate handling bulky culture apparatus and cleaning.

(ii) *CO_2 Incubator:* The animal cell and tissue culture requires a gas phase consisting of O_2 and CO_2 for their growth. This can be done in sealed flasks in a regular dry incubator or hot room. But petri dishes or multiwell plates require a controlled atmosphere with high humidity and elevated CO_2 tension. The cheapest way of controlling the gas phase is to place the cultures in a plastic box, a desiccator or an anaerobic jar. Fill the container with the correct CO_2.

The commercially available CO_2 incubators are more expensive, but their ease of use and the superior control of CO_2 tension and temperature justify the expenditure. The humidity can be controlled by blowing air over a humidifying tray and CO_2 tension can be controlled with CO_2 monitoring device. CO_2 tension can also be controlled by mixing out and CO_2 in the correct ratio.

(iii) *Cell Counter:* A cell counter is used to study the quantitative growth kinetics. It is useful when two or three lines are carried. Sophisticated automated cell counting and size analysis models are manufactured by many companies. For routine counting the "Coulter D Industrial" is more adequate and less expensive.

(iv) *Upright Microscope:* In addition to an inverted microscope, an upright microscope is also required for chromosome analysis, mycoplasma detection and autoradiography. A high grade research microscope such as bright field optic microscope, phase contrast microscope and fluorescence optic microscope with a magnification of 40 X and 100 X are used for mycoplasma testing and fluorescent antibody observation.

(v) *Dissecting Microscope:* The dissection microscope is used to observe small pieces of tissues of embryonic organs and smaller invertebrates. It is also essential for counting of monolayer colonies and also picking up of essential colonies from agar plates.

Apart from the above equipments a fine electronic balance, a simple pH meter for the preparation of media, special reagents are required. A temperature recording thermometer with ranges below −50°C to about +200°C, can be used to monitor ovens, incubators, hot and cold rooms regularly. A magnetic stirrer is essential for rapid stirring action, for dissolving chemicals, disaggregation and suspension culture.

Roller bottles are useful to scale up monolayer and cultures. The right choice of the bottle i.e. the size and number of bottles are calculated from the number of cells required, the maximum attainable cell density and the surface area of the bottles.

It is necessary to use micro pipettes of variable ranges for dispensing small volumes (< 1ml) of reagents. The different ranges of micro-pipettes are 0.5 μl to 1000 μl.

A colony counter is necessary to count monolayer colonies of many plates. Simple plate can be easily counted by eye or dissection microscope.

2.4 Consumable Items

Consumable items includes pipettes and pipette containers, culture flasks, ampoules for freezing centrifuge tubes (10-15 ml, 50 ml, 250 ml) disposable cylinders and needles, filters of various sizes for sterilization of fluids, surgical gloves and paper towels.

The choice of culture vessel is determined by yield required (cell number), whether the cell is grown in monolayer or suspension and the sampling regime i.e. whether the samples can be collected simultaneously or at intervals over a period of time.

Sterile petridishes (9cm) are required for dissection, 5 ml bijoux bottles, 30 ml universal containers, 50 ml samples centrifuge tubes for centrifugation and 1.2 ml plastic vials for freezing in liquid nitrogen.

Syringes are required for filtration is conjuction with syringe filter adapters. They may also be required for extraction of reagents (drugs, antibiotics or radioisotopes) from sealed vials.

Though permanent apparatus is available for sterile filtration, now most of the laboratories are using disposable filters. Common filters being used are 25mm syringe adapters or filter flasks.

QUESTIONS

1. Describe various laboratory requirement for the establishment of animal cell culture.

SHORT QUESTIONS

1. CO_2 incubator (BU April 2000)
2. Cell Counter
3. Coulter Counter (BU Nov. 1997)

3

The Culture Media

3.1 THE SUBSTRATE ON WHICH CELLS GROW

The majority of vertebrate cells cultured *in vitro* are grown as monolayer on an artificial substrate. Hemopoietic cell lines exhibit spontaneous growth in suspension, rodent ascites and tumor cell lines. Most cells need a substrate to proliferate cell which require attachment for growth are said to be anchorage dependent.

Anchorage independent cells can be grown in suspension. The transformed cells are anchorage independent (transformation means phenotypic modification of cells in culture). The cells, which require a substrate may be adhesive (glass, plastic, palladium etc. or a metallic surface; these surfaces may sometimes be variously modified) or own adhesive (agar, agarose, methocel) are discussed below.

(i) *Glass as a substrate:* Slides, coverslips, flasks and test tubes are commonly used glass substrate. It is cheap, easily washable without losing its growth supporting properties and can be sterilized by dry heat (oven) a moist heat (autoclave) and is optically clear for microscopic examination of cultures.

(ii) *Disposable plastic as substrate:* Polystyrene is often most common and cheapest substrate and can be discarded after single use. Polystyrene is hydrophobic and needs to be treated by gamma irradiation, some chemicals or with electric arc to produce a charged surface, which thus becomes wettable.

Other plastics used for tissue culture include polyvinyl chloride (PVC), poly carbonate, polytetrafluoroethylene or teflon (PTFE), thermanox (TPX) etc. PTFE films are available as disposable petri dishes or as membranes to be incorporated in an autoclavable reusable culture vessel. These dishes have two other advantages (1) substrate is permeable to O_2 and CO_2 and (2) plastic is thin and is suitable microscopic sectioning.

(iii) *Palladium and metallic surface as substrate:* Apart from plastic and glass, palladium deposited on agarose can be used as substrate. The growth of fibroblasts and glia. Stainless steel discs and other metallic surfaces can also be used.

3.2 FEEDER LAYER ON SUBSTRATE

Sometimes, on the substrate surface a monolayer of special kind of cells is used for two reasons (1) partly to supplement the culture medium for growth and (2) for conditioning of the substrate by cell products. These monolayers are called "Feeder layers", since they also feed the growing cultures. The feeder layer may consist of mouse embryo fibroblasts, normal fetal intestine, glial cells etc. This feeder layer can effectively be used for selective growth of breast and coelomic epithelium, central and peripheral neurons and also for growing cells used for the production of transgenic animals through transfection.

3.3 THE GAS PHASE FOR TISSUE CULTURE

The cell and tissue culture require a gaseous phase of O_2 and CO_2 for their growth. Cell cultures can grow on an atmospheric oxygen concentration or even at lower O_2 concentration, some organ cultures (eg. late embryo, newborn or adult) may require upto 95% O_2. While O_2 is directly available to cells in cell cultures, in case of organ cultures there may be a problem of diffusion. Since serum free media is in use, O_2 concentration may become an important factor in tissue culture in future.

CO_2 concentration, a complex factor, also needs to be manipulated, since it also influences the pH and the HCO_3^- ion concentration of the culture. At very low concentration of cells, it is necessary to add CO_2 in gas phase of sealed flasks of most cultures. However, in high concentration of cells CO_2 is liberated from the cultured tissue or cells. Therefore, it is not necessary to add CO_2 in gas phase of such sealed flasks with high concentration of cells.

3.4 MEDIA AND SUPPLEMENTS

Although natural media are very useful and convenient for a wide range of uses, they suffer from disadvantages of poor reproducibility due to lack of knowledge of the exact composition. Due to this, synthetic media have been designed for a variety of tissues and purposes like (1) media for immediate survival (2) media for prolonged survival (3) media for indefinite growth (4) media for specialized functions. A balanced salt solution with defined osmotic pressure and pH is needed for immediate survival of cells. Certain inorganic ions, salts and glucose will satisfy the requirement. Serum or the balanced salt solution supplemented with amino acids, oxygen, vitamins and serum protein may be used for longer survival of cells. Eagle (1955) developed a modified basal medium, which is described as "Minimum essential medium"

(MEm) which contains essential amino acids, vitamins and salts. The other more complex synthetic media include the following (1) 199 (Morgan et al., 1950) (2) CMRL 1066 (Parker et al., 1957 (3) RPMI 1640 (Moore et al., 1967) and (4) F12 (Ham's 1965). All these media are usually supplemented with 5-20% serum.

(i) *Balanced Salt medium:* A balanced salt solution (BSS) is composed of inorganic salts consisting of sodium carbonate and most recipes, glucose, although glucose and bicarbonate are often omitted. HEPES buffer (5-20 mM) may be added to these if necessary and equivalent weight of NaCl omitted to maintain the correct osmolarity. BSS is used as solvent for more complete media, as a washing and dissection medium, for short incubation upto 4 hrs, and for variety of other purposes that require an isotonic solution that is not necessarily nutritionally complete.

(ii) *Defined media:* Defined media vary in complexity from Eagle's MEM (Eagle 1959). It consists of large number of different essential and non essential amino acid and vitamins and are often supplemented with extrametabolites (eg. nucleosides, tricarboxylic acid intermediates and lipids) and minerals.

The common constituents of media are as follows:

(a) *Amino acids:* Cultured cells requires essential amino-acids which are not synthesized in the body. Some cells require individual amino acids like cystein and tyrosine. Non essential amino acids are often added to satisfy for particular cell type. The concentration of amino acids added depends upon the maximum cell concentration attainable. The balanced amino acid concentration may influence the cell survival and growth rate. The glutamine is required by most cells. The cultured cells will utilize glutamine as an energy and carbon source (Reitz, et al., 1979).

(b) *Vitamins:* The requirement for entra vitamin is most apparent where the serum concentration is reduced. Most of the vitamins are present in serum.

(c) *Salts:* To maintain the osmoticun in the medium, it is necessary to use various salts. The chiefly available salts are Na^+, K^+, Mg^{2+}, Ca^{2+}, Cl^-, S. Divalent cations particularly Ca^{2+}, Na^+, K^+ and Cl^- regulate membrane potential, while and HCO_3 required for nutritional precursors for macro-molecules, as well as regulators of intracellular charge.

(d) *Glucose:* Glucose is mainly used as energy source in the media. Glucose is metabolized by glycolysis to form pyruvate, which may be converted to lactate or to aceto acetate and enter the citric acid cycle to form CO_2.

(e) *Organic Supplements:* A variety of other compounds including proteins, nucleosides, citric acid cycle intermediates, pyruvate and lipids appear in complex media. These constituents are found to be necessary when the serum concentration is reduced. These may help in cloning and in maintaining certain specialized cells even in the presence of serum.

3.5 SERUM

The most widely used sera in tissue culture are calf, fetal bovine horse and human. Calf serum are being used for culture of more demanding cell lines and for cloning. The human serum is sometimes used in conjunction with some human cell lines. Calf serum is prepared by some workers because it is more consistent from batch to batch. It metabolizes polyamines less due to lower polyamine oxidase. These polyamines are mitogenic for some cells.

(i) *Protein:* Proteins are the major components of serum. The function of these proteins in vitro remain unknown. Relatively few proteins are required for the cell growth and as carriers for minerals, fatty acids and hormones. The proteins which are found to be beneficial are albumins and globulins. Fibronectin (cold-insoluble globulin) promotes cell attachment and α2-macroglobulin inhibits trypsin.

(ii) *Polypeptides:* Natural clot serum stimulates cell proliferation more than serum. This appears to be due to the release of a polypeptide from the platelets during clotting. The polypeptide platelet derived growth factor (PDGF) is one of the family of polypeptides with mitrogenic activity. The PDGF is the major growth factor in serum and it stimulates growth in fibroblasts and glia. Some platlet derived factors like TGF-β may be inhibitory to growth or promote differentiation in epithelial cells.

The other growth factors like fibroblast growth factor (FGF), Epidermal growth factor (EGF), endothelial growth factor and insulin like growth factors IGF-1, IGF-2, have been isolated from whole tissue or released into the medium by cells in culture have varying degree of specificity.

(iii) *Hormones:* The hormones will have different effects on cell and it is difficult to recognize the key pathway. Insulin promotes the uptake of glucose and amino acids. Some growth factors (IGF-1, IGF-2) bind to the insulin receptor but also have their own specific receptors, to which insulin may bind with lower affinity. Growth hormones may be present in sera particularly fetal sera. Hydrocortisone is particularly present in bovine serum in varying amounts. It can promote cell attachment and cell proliferation. In certain condition it can induce cell differentiation.

(iv) *Nutrients and metabolites:* Serum also contains amino acids, glucose, keto acids, lipids, such as oleic acid, ethanolamine and phosphoethanolamine. These may be important in simple media but less important in complex media.

(v) *Inhibitors:* Serum also contain substances inhibiting cell proliferation. These may be bacterial toxins from contamination prior to filtration or the Y-globulin fraction may contain antibodies cross reacting with the culture, but others may be physiological negative growth regulators such as TGF-β. The cytotoxic action of immunoglobins can be reduced by heat inactivation without damaging polypeptide growth factors.

3.6 SERUM FREE MEDIA

Serum free media has been developed due to several disadvantages of serum in culture media. Some of them are:

1. Serum varies from batch to batch and deteriorates within one year.
2. Changing serum batch requires extensive testing.
3. Different cell types require different serum batch, so that a number of batches needs to be maintained and co-culturing may be difficult.
4. The demand of serum usually exceeds the supply for a variety of reasons.
5. The presence of serum is an obstacle to purification when down stream process is used to recover cell culture products.
6. The cost of the serum is high, its cost is ten times the cost of its constituents if used as substitute.
7. Serum may stimulate undesirable growth and may even inhibit growth in some cases.

Since each cell type requires a different recipe, the major advantage of serum free media is, the ability to make a medium selective for a particular cell type. The removal of serum requires (1) high degree of purity of reagents and water (2) high degree of cleanliness of all apparatus. This is necessary, since removal of serum also removes the protective and destroying action, which some of the serum proteins have.

Growth is often slower in serum free media. Fewer generations are achieved with finite cell lines. Some selection procedures are required to adopt cell lines to serum free media.

Finally and often important is the availability of serum free media, subject to proper quality control which is quite limited and the products are often expensive.

3.6.1 Preparation of Serum free media

A number of recipes are now available for particular cell types. Some of these are available commercially, others are to be prepared. The procedure for making up serum free recipes is similar to regular media. Ultra pure reagents and water should be used and care should be taken with solutions of Ca^{2+} and Fe^{2+} or Fe^{3+} to avoid precipitation. Metals salts tend to precipitate in alkaline pH in the presence of phosphate, particularly when autoclaved, so cations in stock solutions should be kept at a low pH (below 6.5) and phosphate free. They should be sterilized by autoclaving or filtration. It is often recommended that these be added immediately before use. Generally the constituents are prepared as a series of stock solutions. The tyrosine, tryptophan and phenylalanine is dissolved in 0.1 N HCl at 50X, essential amino acids are dissolved in water at 100X, salts in water at 10X, minerals and vitamins 1000X in water. Lipids and other special cofactors are dissolved in the appropriate solvents. These stock solutions are mixed in the correct proportion and diluted to the final concentration and the pH and osmolarity checked.

The growth factors, hormones and cell adhesion factors are added separately just before use.

(i) *Commercially Available Serum free media:* Several suppliers now make serum free media (ICN/Flow, GIBCO, clonetics etc.). These medias are primarily designed for culture of hybridomas, commercial recipes are often a trade secret, so it is better to screen a number of media over several subcultures of different cell types.

3.7 NATURAL MEDIA

Natural media is the most preferred medium, since it is the cheapest and most convenient.

The natural media which promote the cell growth fall into the following three categories.

1. Coagula such as plasma clots
2. Biological fluids such as serum
3. Tissue extracts, which embryo extract is the most common.

(i) *Plasma clots:* Plasma clots have been in use since long time. Plasma clots are now available commercially either as liquid plasma in siliconed ampules or as lyophilized plasma. This can be reconstituted by the addition and distilled water saturated with CO_2. One can also prepare plasma from a male fowl, where efforts are made to prevent coagulation of blood by adding heparin as an anticoagulent.

(ii) *Biological fluids:* The most commonly used biological fluid is serum, which is commonly obtained from human adult blood, placental cord blood, horse blood or calf blood. Out of these, human placental-cord serum and fetal calf serum seems to be satisfactory. The serum can be obtained as exuded liquid from blood undergoing coagulation and is filtered using millipore filters. The serum should be tested for sterility and toxicity before using and stored at low temperature. The presence of contaminants can be checked by incorporating the serum in a growth medium. The toxicity of sera can be reduced by heat inactivation.

Besides serum, following are the other biological fluids used as natural media.

(1) amniotic fluid (2) ascitic and pleural fluid (3) aqueous humour (from eye) (4) serum ultrafiltrate (5) insect haemolymph (6) coconut milk (usually used for plant tissue culture).

(iii) *Tissue extracts:* Embryo extract is the most commonly used tissue extract. It is a crude homogenate of 10 days chick embryo clarified by centrifugation. Cohn (1966) fractionated crude extract to give high and low molecular weight fractions. The low molecular weight fraction promoted cell proliferation, while the high molecular weight fraction promoted pigment and cartilage cell differentiation. These fractions were not fully characterized, but recent evidences suggest that the low molecular weight fraction probably contained peptide growth factors and the high molecular weight fraction proteoglycans and other matrix constituents. Smith and Schroedl (1992) showed that embryo extract can be replaced by hemin in the induction of skeletal muscle differentiation.

QUESTIONS

1. Outline the significance of serum, chemically defined and natural media in animal cell culture.
2. "Serum still retains its importance in the preparation of Media". Discuss. (BU May 1996)
3. What are serum free media? What are the advantages and disadvantages of such media? Discuss. (BU Nov. 1999)
4. List out the constituent of serum and add a note on their importance in in-vitro cell culture. (BU April 2001)

SHORT QUESTIONS

1. Serum Free Media (BU Nov. 1999)
2. Natural Media
3. Growth factors (BU Nov. 1997)

4

Primary Culture and Disaggregation of the Tissue

There are two methods of obtaining a primary cell culture.

1. The cells which are adhering to a substrate are allowed to migrate out from the tissue.

2. A suspension of cells can be obtained by disaggregating the tissue either mechanically or enzymatically. Most normal untransformed cells survive and proliferate to produce primary culture when attached to a substrate and these cells can be obtained by disaggregation. The disaggregated cells can also grow in suspension to produce the primary culture.

The enzymatic disaggregation of cells causes less cellular damage and the yield is also comparatively more than that obtained from mechanical method. The enzymes usually used are crude preparations of trypsin, collagenase, elastage, hyaluronidase, DNase, pronase, and dispase. Various combinations of trypsin and pronase gives the most complete disaggregation but may damage the cells. Collagenase and dispase gives incomplete disaggregation but are less harmful. Hlyaluronidase in combination with collagenase can be used to digest intracellular matrix. The DNase is used to disperse DNA release from lysed cells, as it tends to impair proteolysis and promote reaggregation.

In the mechanical method, the tissue is forced through cheese or silk cloth or shaking the tissue with the glass beads in an appropriate buffer. The main disadvantage of this method is that it leads to certain degree of cell damage and also low yield.

4.1 SOURCE OF TISSUE

The source of tissue from which the cells are isolated may be embryogenic/

adult, normal or neoplastic cells. Cultures from embryogenic tissue survive and proliferate better than those cultured from the adult. This is due to the low level of specialization and higher proliferation capacity in embryos. Adult tissue have lower growth rate and consists of non replicating specialized cells. These tissues may be stored in a refrigerator before transferring it to BSS or to a culture medium because viable cells may be recovered from chilled tissue several days after explanation.

The excised tissue is sterilized with 70% alcohol and the tissue is removed aseptically and transferred to a balanced salt solution (BSS) or to a suitable culture medium.

4.1.1 Isolation of Mouse Embryo Tissue

(1) Uterus is removed aseptically from the pregnant mouse, and the embryo is dissected. The optimal age for preparing cultures from the whole disaggregated embryo is around 13 days, when the embryo is relatively large and contains a high proportion of undifferentiated mesenchyme. This is essential because most of the cultures are derived from mensenchyme.

(2) Embryos are dissected from the intact uteri in the tissue culture laboratory. Uterus points are torn and the embryo is removed carefully. As the uterus is torn apart, the embryos are set free from the membrane and placenta. These embryos are allowed to bleed in petridish and then transferred to a fresh petridish.

(3) The dissected embryos are taken in a 25 ml or 50 ml screw capped tube containing 10 to 20 ml of BSS. Antibiotics can be added to the BSS to avoid risk of infection.

4.1.2 Isolation of Embryo from Hen's egg

(1) The 8 day old hen's egg can be used for the culture of dispersed cells from the whole embryo. The method of isolation of embryo from hen's egg is shown in the Fig. 4 (1a-e).

(a) Swab the egg with 70% alcohol. Crash the top of the shell and peel off to the edge of the air sac with sterile forceps. Resterilize the forceps with alcohol and peel of the white shell membrane to reveal the chorioallantoic membrane (CAM) below, with its blood vessel.

(e) Pierce the CAM with sterile curved forceps and lift out the embryo by grasping gently under the head.

(f) Transfer the embryo to a dish containing BSS.

4.1.3 Isolation of Human Biopsy material

(1) Certain problems are associated with the handling of human biopsy material. To use biopsy material, it is necessary to obtain a consent from the hospital ethical committee, from the patient and physician or surgeon who is attending the patient. further more, biopsy sampling is usually performed for

diagnostic purposes and hence it is necessary to meet the pathologist first.

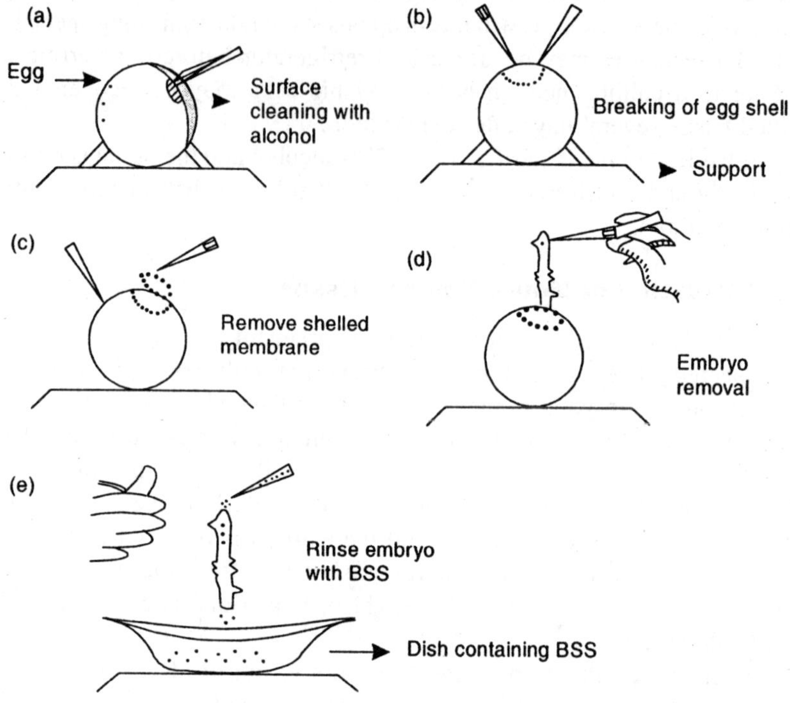

Fig. 4.1. Methods of chick embryo isolation

(2) Biopsy material has to be handled in a class II biohazard cabinet as it carries a risk of infection. All media and apparatus is disinfected by autoclaving or immersion in suitable disinfectant.

(3) After surgery, a sample is collected and transferred to tissue culture laboratory. The tissue is stored at 4°C. Biopsy sample will survive at least for 24 hrs and even upto 3 or 4 days. Longer storage may cause deterioration of tissue.

4.2 PRIMARY CULTURE

4.2.1 Primary Explant Technique

Harrison (1907) & Corel (1912) were the first persons to develop a primary explant technique. In this technique, a fragment of tissue was embedded in blood plasma or lymph mixed with embryo extract and serum and placed on a slide. The clotted plasma holds the tissue in place. Both the embryo extract and serum will serve as nutrient supplies and stimulate the migration of the explant across the solid substrate. The heterologous serum was used to promote clotting of the plasma. The simplified technique for primary explant preparation (See Fig. 4.2) is described as follows.

(1) The tissue is chopped finely, rinsed and placed on to a medium with a high concentration (40-50%) of serum.

(2) Transfer tissue to fresh sterile BSS and rinse, dissect off unwanted tissue like fat or neurotic material and chop finely with crossed scalpels to about 1mm cubes.

(3) Wash the chopped tissue in BSS and allow them to settle then remove the supernatant fluid two or three times.

(4) Transfer the pieces to a culture flask containing 1 ml growth medium per 25 cm^2 growth surface. Cap the flask and place in incubator or hot room at 36.5°C for 18-24 hr.

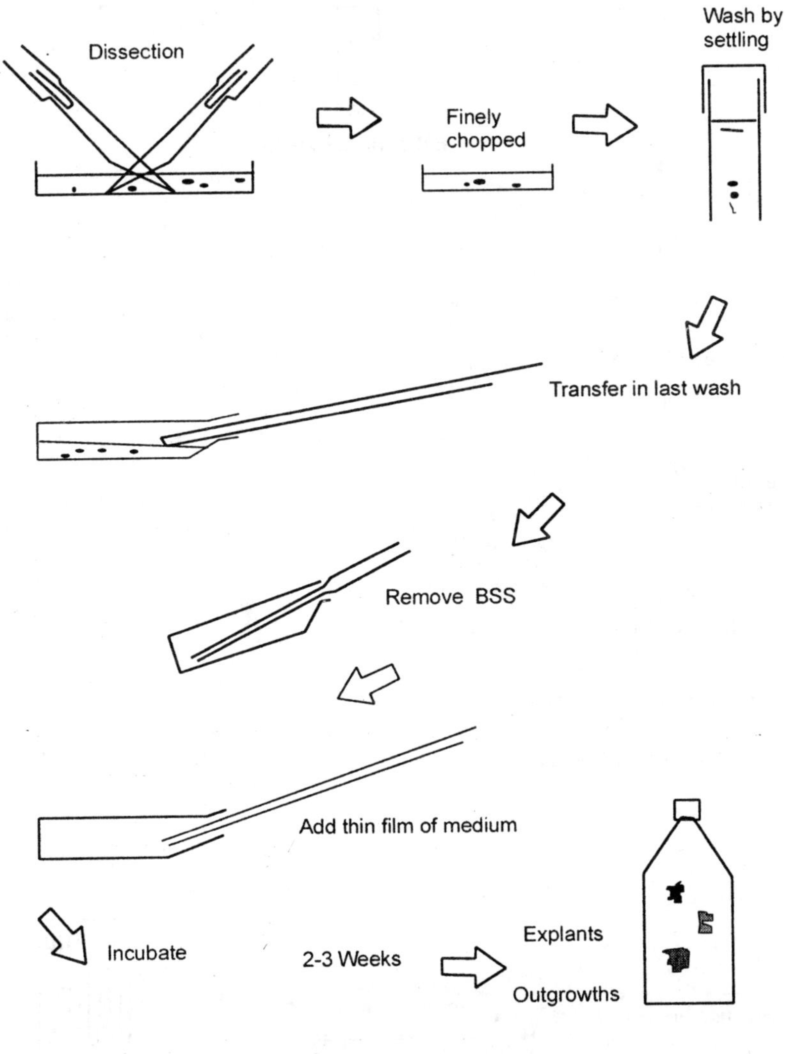

Fig. 4.2. Primary explant technique

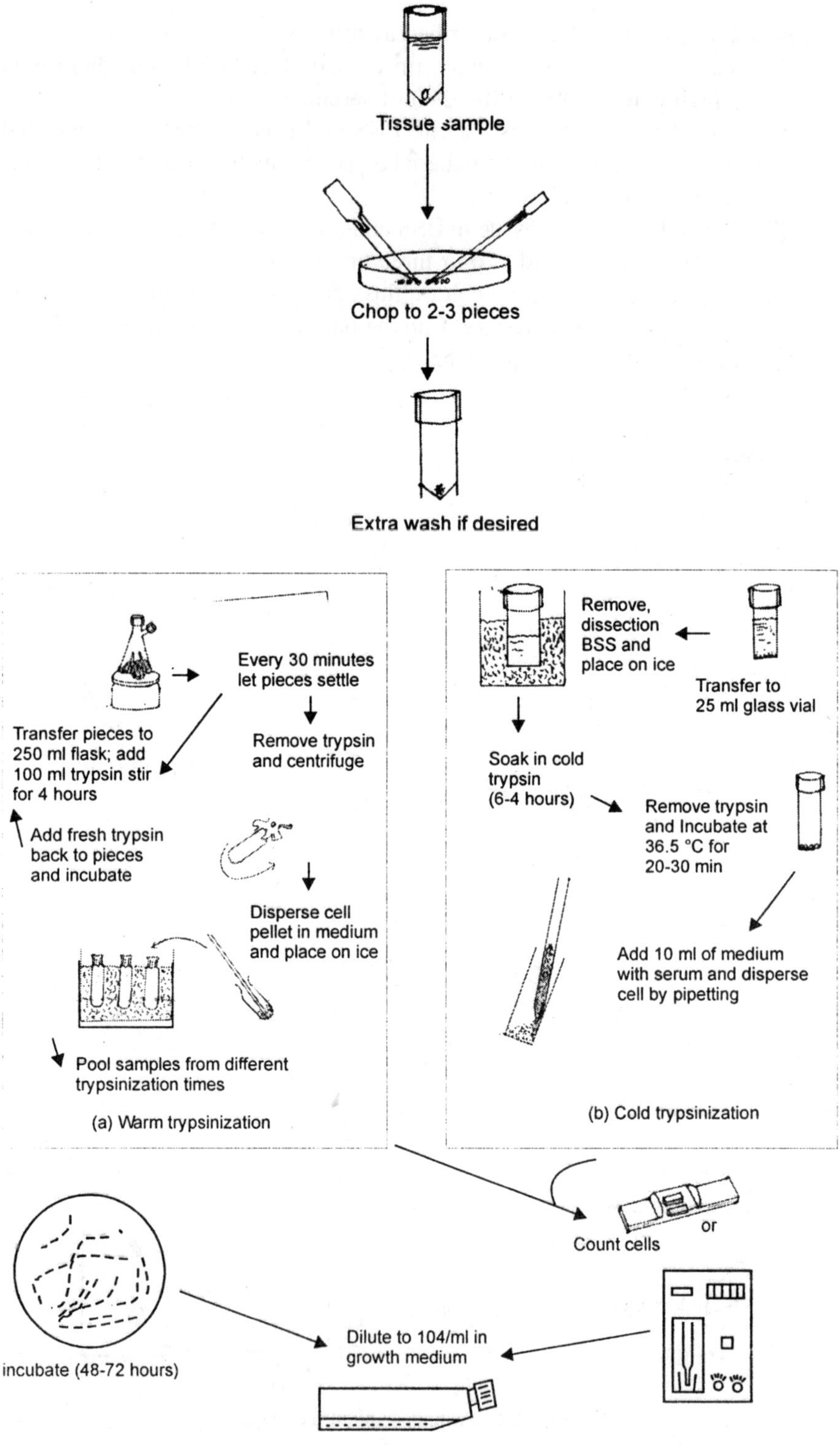

Fig. 4.3. Preparation of primary culture by disaggregation using trypsin. (Warm trypsin method on the left and cold trypsin method on the right

This technique is particularly useful for small amounts of tissue such as skin biopsies where there is a risk of losing cells during mechanical or enzymatic disaggregation.

4.2.2 Enzymatic Disaggregation

Enzymatic or mechanical disaggregation gives a much higher yield of cells in a short duration when compared to primary explant technique.

Embryonic tissue disperses more readily and gives a higher yield of proliferating cells than new born or adult. This is because in the latter case, there are fewer undifferentiated cells and more fibrous connective tissue and extra cellular matrix. In several tumors, the tissue is mostly destroyed during enzyme treatment and viable cells are obtained with difficulty.

(a) *Disaggregation by trypsin:* The crude trypsin is the most common enzyme used for disaggregation of embryonic tissue. The method of disaggregation is described as follows:

(1) The explant is exposed to the warm enzyme (36.5°C) for a minimum period and the dissociated cells are collected every half an hour.

(2) It will take 3-4 hours for the complete disaggregation of tissue. Then trypsin is removed by centrifugation.

(3) Alternatively cold trypsinization can be done which involves soaking of tissue in trypsin at 4°C to allow penetration of enzyme, followed by incubation at 36.5°C for a shorter period.

Different steps of warm trypsinization and cold trypsinization are shown in Fig. 4.3.

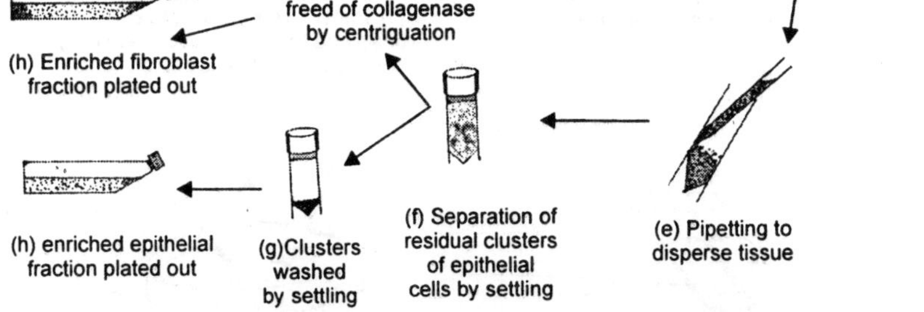

Fig. 4.4. Stages in disaggregation of tissue for primary culture by collagenase

(b) *Disaggregation by Collagenase:* Collagenase is used for disaggregation of embryonic, normal and malignant tissues. Disaggregation by trypsin causes damage to epithelial cells and they are ineffective on fibrous tissues. The collagenase is effective for disaggregation of intracellular matrix, which are rich in collagen. The method of disaggregation by collagenase is as follows:

(1) 0.5 ml of crude collagenase is added to a finely chopped tissue in complete medium.

(2) It is then incubated at 36.5°C for 4-48 hr. without agitation disaggregation is slow in case of tumor tissues. In that case it is left in the incubator upto 5 days.

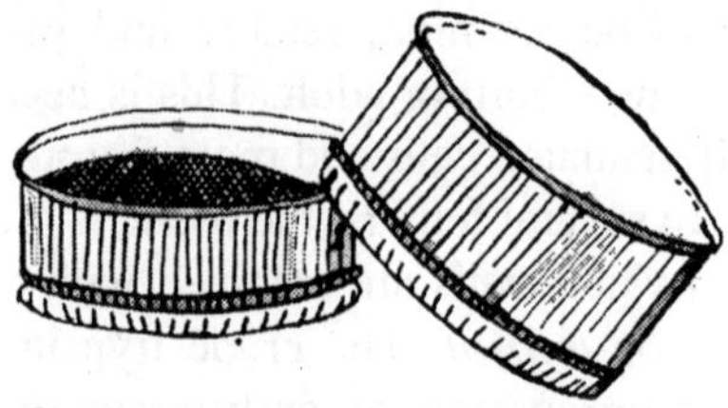

(a) Stainless steel sieves for disaggregation

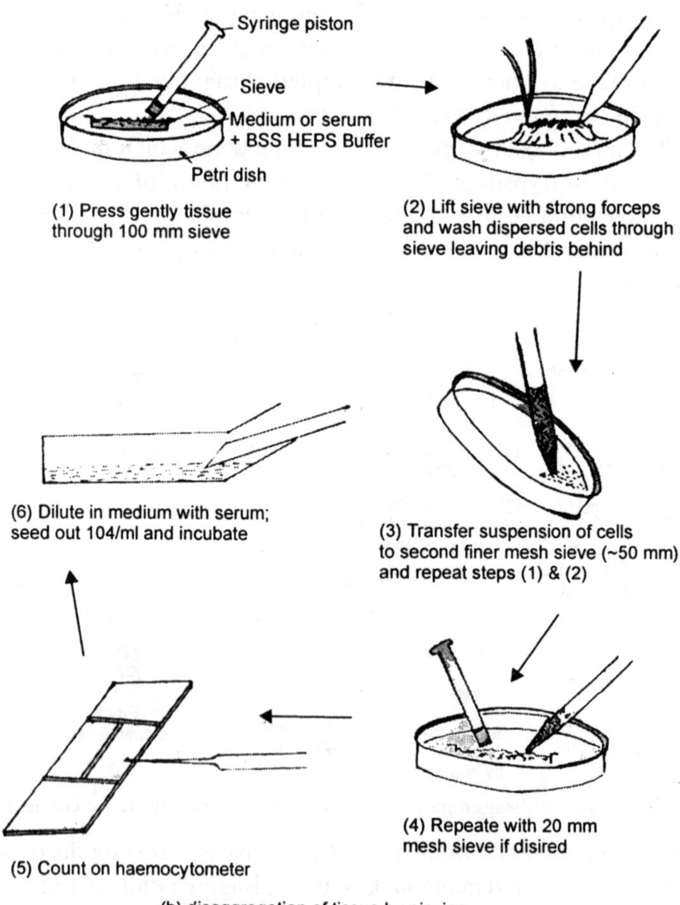

(b) disaggregation of tissue by sieving

Fig. 4.5. Different steps involved in disaggregation of tissue by sieving

(3) After disaggregation, collagen is removed by centrifugation and cells can be seeded and cultured at a high concentration.

The different steps involved in disaggregation of the cells by collagenase are shown in the Fig. 4.4.

4.2.3 Mechanical disaggregation

Enzymatic digestion of tissue involves intense labour and a risk of cells getting damaged during the process. An alternative method of disaggregation of cells by the mechanical means. Steps involved in this procedure are:

(1) The tissue is carefully sliced and the cells are collected and placed few at a time into a stainless steel sieve of 1mm mesh in a 9 cm petridish.

(2) Force the tissue through the mesh into the medium by applying gentle pressure using the piston of a disposable plastic syringe or even repeatedly pipetted.

(3) Pipette the partially disaggregated tissue from the petri dish into a sieve of finer porosity.

The suspension may be diluted and cultured in a suitable medium.

Though this method causes mechanical damage to the cells, the suspension is more quickly obtained than by the enzymatic disaggregation. By this method only a good yield of cells in a shorter duration can be obtained but not its efficiency. The steps involved in the mechanical method are shown in the Fig. 5.5.

4.2.4 Separation of viable and non viable cell

The dissociated cells obtained as above are described as primary cells. These cells grow well when seeded on culture plates at high density and culture can also maintained in suspension. In case of adherent culture, the non viable cells can be removed by initial change of medium, or centrifugation of the cells in a mixture of ficoll and sodium metrizoate. In this gradient centrifugation viable cells are collected from the interface.

QUESTIONS

1. What is primary culture? Explain any two methods employed for tissue disaggregation. (BU April 2001)
2. Describe the efficacy of Enzymatic and Mechanical disaggretation of tissue.

SHORT QUESTIONS

1. Primary culture
2. Isolation of embryo

5

Differentiation of Cells and Growth Kinetics

The differentiated cells lose their phenotypic properties when cells are cultured and propagated as a cell line *in vitro*. Their characteristic features will be different from the tissue from which it was derived. This is due to several factors, many of them which are yet undefined and they regulate the geometry, growth and function *in vito*, but are absent from the tissue culture environment.

Differentiation is the process of expression of phenotypic properties characteristic of the functionally mature cell *in vivo*. "Differentiation" here is described as the combination of constitutive and adaptive properties found in the mature cell.

"Differentiation" is described as the loss of the differentiated properties of a tissue when it becomes malignant or when it is grown in culture. The loss of differentiation in a cell is due to the following factors: cell death, selective over growth, adaptive responses and loss of certain phenotypic properties.

5.1 STAGES OF COMMITMENT AND DIFFERENTIATION

There are two main pathways for differentiation in an adult organism. (1) A small population of totipotent or pluripotent undifferentiated stem cells gives rise to committed precursor cells. These will progress toward terminal differentiation and loose their capacity to divide as they reach the terminal stages. This gives rise to a fully mature, undivided differentiated cells. (2) Fibrocyte cells fail to differentiate when their local reduction in cell density or the presence of one or more growth factors. When tissue has regained the appropriate cell density, cell proliferation stops and differentiation is reinduced.

The first one is used for hemopoietic system, the skin and the gastric mucosa, while second is used for regeneration and is not continual and requires quick response to trauma, such as in wound repair or liver regeneration.

5.2 PROLIFERATION AND DIFFERENTIATION

As the differentiation progresses, the cell division is reduced and is eventually lost in most of the cell system. However, tumor cells will continue to synthesize melanin when the cells are proliferating. Generally differentiated product synthesis will increase when the cell division stops.

Majority of the cell lines do not express fully differentiated properties, in such cases, it is difficult to analyse the cells for proper biochemical or molecular analysis. Hence, by providing correct environment it is possible to reinduce differentiated phenotype in pure population of cells.

5.2.1 Markers of Differentiation

Marker differentiation allows differentiation of cell properties and regulates its expression. Marker which is expressed early and retained throughout the subsequent maturation stages are generally regarded as 'lineage markers' eg. cytokeratin protein and glial fibribllary acidic protein (astrocytes) eg. markers representing terminal differentiation. Enzymes involved in the synthesis of these products e.g hemoglobin in an erythrocyte, serum albumin in a hepatocyte, transglutaminase in a differentiating squame or glycerol phosphate dehydrogenase in an oligodendrocyte. These properties are often expressed well, these are working under adaptive control by hormones.

Differentiation should be regarded as the expression of one or more than one of these marker properties. The lineage markes are helpful in conforming cell identity and expression and the functional properties of the mature cells.

5.3 INDUCTION OF DIFFERENTIATION

There are four main parameters governing the control of differentiation:

5.3.1 Soluble inducers

Soluble inducers include the endocrine hormones such as hydrocortisone, glucagon and thyroxine (or triidotyrosine), paracrine factors released by one cell and influencing adjacent cells, which are poorly characterized (eg. TGFβ from platelets, prostaglandins, NGF, glia maturation factor), alveolar maturation factor and interferons, vitamins such as vitamin D and retinoic acid and inorganic ions, particularly Ca^{2+}, where high Ca^{2+} promotes keratinocyte differentiation.

5.3.2 Nonphysiological inducers

Friend in 1967 observed that mouse erythroleukemia cells treated with DMSO turned red due to the production of hemoglobin. Later on series of experiments proved that many other cells like neuroblastoma, myeloma, and mammary carcinoma also responded to DMSO by differentiation. The action of compounds like hexamethylene bisacetamide, N-methyl acetamide, sodium butyrate, benzodiazepines etc. can be related to that of DMSO.

The action of these compounds are yet concluded but may be mediated by changes in membrane fluidity (particularly the polar solvents like DMSO and the anaesthetics and tranquilizers), by their influence as lipid intercolators on enzymes of signal transduction such as protein kinase C (PKC) or phospholipase D (PLD) or by alterations in DNA methylation or histone acetylation.

Tumor promoters such as phorbol meristate acetate (PMA) have been shown to induce squamous differentiation in bronchial mucosa.

5.3.3 Cell interaction

Homologous cell interaction occurs optimally at high cell density and it involves gap junctional communication, where metabolites, second messengers such as cyclic AMP, diacylglycerol (DAG), Ca^{2+} or electrical charge may be communicated between cells. This interaction probably hormonizes the expression of differentiation within the population of similar cells, instead of initiating its expression.

Heterologous cell interaction occurs between mesodermally and endodermally or ectodermally derived cells. This is responsible for initiating and promoting differentiation. Mutual interaction between cells originating in different germ layers promotes differentiation during gastrulation in the embryo and latter during organogenesis.

Hepatocyte growth factor (HGF), which belongs to the family of heparin-binding growth factors (HBGF), has been shown to be released from fibroblasts, such as the MRC-5, and it induces tubule formation in the MDGK continuous cell line from dog kidney; KGF (another HBGF) is produced by dermal fibroblast and influences epidermal differentiation. Both factors are made by fibroblasts and are prime candidates for paracrine differentiation factors.

5.3.4 Cell-matrix interactions

The surrounding surface of most cells consists of a complex mixture of glycoproteins and proteoglycans. these are highly specific for each tissue and even parts of a tissue. Reid (1990) has shown that construction of artificial matrices from different constituents can regulate gene expression. Addition of liver derived matrix material will induce the expression of the albumin gene in hepatocytes. Collagen has been found to be essential for the functional expression of many epithelial cells and for endothelium to mature into capillaries.

The RGD motif (arginine-glycine-valine) in matrix molecule appears to be the receptor interactive moiety in many cases. Small polypeptide containing this sequence will effectively block matrix-induced differentiation implying that the intact matrix molecule is required.

Endothelium and many epithelial cell types are proved to be effectively differentiate in martigel, a matrix material produced by the Engelberth Holm Swarm (EHS) sarcoma, made up of predominently, laminin but also contains collagen and proteoglycans. This is a useful technique but has the problem of introducing another biological variable to the system.

The extracellular matrix also plays a role in the modulation of growth factor activity. It has been suggested that the matrix proteoglycans, particularly heparan sulfate proteoglycans (HSPGs), may bind certain growth factors, such as GM-CSF, and make them more available to adjacent cells.

5.4 DIFFERENTIATION & MALIGNANCY

It is frequently observed that increasing progression of cancer, histology of the tumor indicates poorer differentiation and these poorly differentiated tumors will have a lower survival rate than those with differentiated tumors. It has been stated that, cancer is, principally a failure of cells to differentiate normally. Therefore, it is surprising to find that many tumors grown in tissue culture can be induced to differentiate. Much of the fundamental data on cellular differentiation has been derived from the friend murine leukemia, mouse and human myeloma, hepatoma and neuroblastoma. Nevertheless, there appears to be an inverse relationship between the expression of differentiated and malignancy-associated properties.

Tumor cells often retain the ability to respond to differentiation inducers. Friend erythroleukemia responds to DMSO but not to erythropoietin, while normal bone marrow and fetal mouse liver respond to erythropoietin, but there are no reports of induction by DMSO.

Tumor cells, or at least a proportion of them respond to differentiation induction, they may not do so *in vivo*. The reasons for this are complex, (1) over expression of the 'growth mode phenotype' due to over expression of oncogenes and reduced or abnormal expression of tumor suppressor genes still expressed *in vitro* (2) part of the reason may be the failure of normal cell-cell and cell-matrix interaction. The matrix is modified in such way that inability of the tumor cell should produce correct constituents and increased production of extra cellular proteases and glycosidases, which further modifies the matrix.

The tumor cells are useful models for the study of differentiation and the production of specialized products.

5.5 GROWTH KINETICS

The culture cycle beginning with a lag phase, proceeds through the logarithamic phase to a stationary phase and finally to the decline and death of cells. Although cell growth usually implies increase in cell numbers, increase in cell mass can occur without any replication. The difference in mean cell mass between cell population is considerable, as would be expected, but so is the variation within the same population.

Growth can be defined in the following term:

(a) Specific growth rate, μ (i.e. the rate of growth per unit amount (weight/numbers of biomass):

$$\mu = (1/x)(dx/dt)\,h^{-1}$$

where dx is the increase in cell mass, dt is the time interval and X is the cell mass. If the growth rate is constant (e.g. during logarithmic growth), then

$$\text{In } X = \text{In } X_0 = \mu t$$

where X_0 is the biomass at time t_0

(b) Doubling time td (i.e. the time for a population to double in number/mass):

$$t_d = \frac{\text{In}_2}{\mu} = \frac{0.693}{\mu}$$

(c) Degree of multiplication n or number of doubling (i.e. the number of times the inoculum has replicated):

$$n = 3.32 \log (X/X_0)$$

QUESTIONS

1. Describe cell differentiation? Add a note on factors that induce cell differentiation.

(BU Nov. 1999)

SHORT QUESTIONS

1. Growth kinetics

6

Animal Cell Line and their Origin and Characterisation

In order to maintain the cell line and proliferation of the primary culture produced on a substrate or in suspension medium, the primary culture needs to be subcultured. Hence cell proliferation has become an important feature. The primary culture may have variable growth fraction, depending on the type of cells present in culture, after the first sub-culture the growth fraction is usually high (80% or more).

Containing many types of cells derived from tissue, during subculturing (passages or transfer) a more homogeneous cell line can be obtained by a series of subculture of cells, derived from the primary culture from a very heterogeneous primary culture of the original tissue. The culture now called a cell line, can be propagated, characterised and stored. The potential increase in cell number and uniformity of cells opens up a much wider range of possibilities. The term "cell line" implies the presence of several cell lineages either similar or distinct. Among these cell lineages, a particular cell lineage with specific properties, are identified in the bulk of cells of that culture, which is described as a *"cell strain"*. A cell line or cell strain may be finite or continuous depending on whether it has limited culture-life span or it is immortal in culture. Finite cell lines grow upto 20-80 population, doubling before extinction. Some commonly used cell line and cell strains are listed in Table (3) and a comparison of the properties of finite and continuous cell lines is given in Table (4).

Every cell line is given a code or designation in order to differentiate from the several other cell lines used from the same source. For eg. normal human brain (NHB), a cell strain or cell number NHB_1, NHB_2, etc. and if cloned a clone number NHB_{2-1}, NHB_{2-2} etc.

For finite cell lines the number of population doubling should be estimated and indicated by a slash number eg. NHB2/2 and with increase by one for a split ratio of 1 : 2 (NHB2/2, NHB2/3 etc), by two for a split ratio of 1 : 4 (NHB2/2, NHB2/4 etc) and so on. This is more difficult with a continuous cell line, where the generation number is very high and often known.

6.1 MAINTENANCE OF ADHERENT CELL LINES:

Most of the monolayer primary cultures or continuous cell lines need a periodic change of medium (i.e. removing old medium and adding fresh medium), whether or not the cells are proliferating. In cultures, where cells are proliferating, the usual practice in subculturing an adherent cell line involves the following steps.

(1) Removal of the medium and add PBSA along the side of the flask opposite the cells, to avoid dislodging of cells, rinse the cells and discord the PBSA. This is designed to remove traces of serum that would inhibit the action of trypsin.

(2) Dissociation of the cells in the monolayer with trypsin or other enzyme.

(3) Incubate until cells round up (this takes after 5-15 min.), when the bottle is tilted, the monolayer should slide down the surface.

(4) Add medium (0.1-0.2 ml/cm^2) and disperse cells by repeated pipetting over the surface bearing the monolayer. To disperse the cell line, pipette the cell suspension up and down for few minutes to get single cell suspension.

A single cell suspension is desirable at subculture to ensure an accurate cell count and uniform growth. It is necessary, where cells are to be isolated as clones.

Intervals between change of medium and subculturing vary from one cell line to another depending upon the rate of growth of metabolism. Rapidly growing cell lines like HeLa are subcultured once per week and the medium changed every four days. More slowly growing cell lines are subcultured every 2, 3 or 4 weeks and the medium is changed every week between subcultures.

There are several factors which are used to assess the need for replacement of medium. These include (1) drop in pH (2) Cell contraction (3) Cell type (4) Cell morphology etc.

6.2 MAINTENANCE OF CELL LINES IN SUSPENSION:

The primary cultures or continuous lines are continuously grown in suspension because of two reasons:

(1) they are nonadhesive (e.g. many leukemias and murine ascites tumors)

(2) they are kept in suspension mechanically and can be subcultured like microorganisms. Trypsin treatment is not required and the whole process is quicker and less traumatic for the cells. Subculturing is done by dilution of cell density, rather than by replacement of the medium, since latter will require centrifugation of the suspension

culture to obtain a pellet of cells to be added to a fresh quantity of the medium. Following protocol is being used.

(1) Mix cell suspension followed by disperse to avoid lumps.

(2) The cell samples is removed for counting.

(3) Add sufficient cells to give final concentration of 10^5 cells/ml for slow growing cells (24-48 hr. doubling time) or 2×10^4/ml for rapidly growing cells (12-18 hr. doubling time).

(4) Cap the flask and incubate the culture.

Suspension cultures have a number of advantages. The production and harvesting of large quantities of cell may be achieved without increasing the surface area of the substrate.

Monolayer cultures are convenient for cytological and immunological observations, cloning mitotic "shake off" (for cell synchronization of chromosome preparation) and *is situ* extraction without centrifugation.

6.2.1 Characterization of Cell Line

The characterization of various cell lines is required for the following reasons.

(1) Correlation with the tissue of origin i.e. to identify the lineage to which the cell belongs to understand the position of cells within in a lineage and to check whether the cells have been transformed or not.

(2) To maintain the genetic instability and phenotypic variation.

(3) To check cross contamination and conformation species of origin.

(4) Identification of specific cell lines within a group from the same origin. Selected cell strain or hybrid cell lines, all of which require demonstration of features unique to that cell line or cell strain.

Following are the some of the technique to characterise cell lines.

6.2.1.1 'Chromosome Banding' (Karyotyping)

This technique helps the identification of individual chromosome pairs. when there is little morphological difference between them.

Eg. Giemsa banding also called G-banding, Q-banding C-banding.

6.2.1.2 Chromosome painting

Fluorescently labelled probes that will bind to specific genes on the chromosome, it has become possible to locate genes, identifying translocation and determine the species of origin of chromosome. This technique employs *in situ* hybridisation technology which can also be used for extrachromosomal and cytoplasmic localisation of specific nucleic acid and can be used to localise specific mRNA species.

6.2.1.3 DNA Content

The amount of DNA per cell is relatively stable in normal cell lines, such as human fibroblast, chick and hamster fibroblast, but varies in cell lines from the mouse and from many neoplasms. The content of DNA can be analysed using flow-cytometer, microfluorometry, microdensitometry, ethidium bromide fluorescence etc. DNA content analysis is useful in characterisation of transformed cells that are often aneuploid and heteroploid.

6.2.1.4 DNA Hybridisation

The extracted DNA is cleaved with restriction enzymes, electrophoresed, blotted on nitrocellulose membrane and hybridised with labelled probes. This provides information about species-specific regions or amplified regions of the DNA that are characteristic to that cell line. Thus strain-specific gene amplifications, such as amplification of the dehydrofolate reductase (DHFR) gene, may be detected in cell lines selected for resistance to methotrexate.

6.2.1.5 DNA Finger Printing

Since DNA contains region known as satellite DNA, these are apparently not transcribed and these will give rise to regions of hypervariability. When the DNA is cut with specific endonucleases, it can be probed with cDNA's that hybridise to these hypervariable regions. Electrophoresis reveals the variations in fragment length that are specific to individual from which the DNA was derived. When analysed by PAGE, each individual's DNA gives a specific hybridisation pattern. With this technique, the origin of a cell line and any cross contamination can be confirmed.

6.2.1.6 Iso Enzymes

The iso enzymes can be used to characterise cell lines, due to polymorphisms among species and sometimes among races or individuals within the species. The isoenzymes can be separated chromatograpically or electrophoretically and distribution patterns (zymograms) of species or tissue can be found. Paul and Fottrell (1961) demonstrated differences in esterase zymograms between normal and malignant mouse cells and human cells.

6.3 USE OF MARKERS

Marker are used to characterise and identifying a cell line. These are unique and specific to cell lines.

Eg. Cell surface antigens, immediate filament proteins, differentiated products such as haemoglobin, melanin, myosin etc.

6.3.1 Marker Gene

It is also called reporter gene. These genes produces a phenotype which is easily and specifically detected or which allows a differential multiplication of the cells.

Examples: Thymidine kinase, DHFR, CDA protein, HGPRT, Neomycin phosphotransferase.

(a) *Thymidine kinase (TK)*: Nucleotide synthesis through salvage pathway requires the activity of thymidine kinase (TK). It is necessary in the synthesis of TTP, it phosphorylates thymidine. TK deficient (TK^-) cell are killed on HAT medium, which contains DNA nucleosides, hypoxanthine thymidine and aminopterin.

Aminopterin blocks the *De-novo* (endogenous) pathway of nucleotide synthesis. The cell are forced to follow salvage pathway for their survival. TK^- cells are unable to use thymidine present in the medium for nucleotide starvation on HAT medium which acts as an extremely efficient selection agent for TK^+ cells.

The TK genes can be used as selectable marker only when TK^- hosts cells are used for transfection. The transfected cell are cultured on the HAT medium, on which only TK^+ cells can survive and multiply.

(b) *Dihydrofolate Reductase*: (methotrexate resistance) (DFFR) normal cells are sensitive to methotrexate, even at very low concentration (0.1mg/ml). Methotrexate is an inhibitor of DHFR enzyme, involved in the biosynthesis of nucleotides by *de-novo* pathway.

Cell lines resistant to methotrexate have been developed. This can be exploited as marker gene for cell line selection.

(c) *Neomycin phosphotransferase*: (G 418 Resistance) G 418 is an aminoglycoside antibiotic which inhibits protein synthesis. Mammalian cells are sensitive to G 418. Neomycin phosphotransferase confer resistance to G 418. Various constructs (plasmids/cosmids) of neomycin phospho transferase gene have been prepared and used for transfection of a variety of mammalian cell lines. Transfected cells are efficiently selected in a medium containing G 418.

(d) *CDA proteins (PALA Resistance)*: Normal mammalian cells are sensitive to N-phosphoacetyl-L-aspartate (PALA), an inhibitor of CDA protein required for uridine biosynthesis. CDA gene is isolated from syrian hamster. Transfected mammalian cells are highly resistant to PALA. This can be exploited as a marker gene for selection of cell lines.

Table 3. Cell lines and cell strains in regular use and their origin

Name	Morphology	Origin	Age	Tissue	Ploidy	Characteristic
1. *Finite cell lines*						
MRC 5	Fibroblast	Human lung	Embryonic	Nor	Diploidy	Susceptible to human viral infection
MRC 9	Fibroblast	Human lung	Embryonic	Nor	Diploidy	Susceptible to human viral infection
W 138	Fibroblast	Human lung	Embryonic	Nor	Diploidy	Susceptible to human viral infection
IMR 90	Fibroblast	Human lung	Embryonic	Nor	Diploidy	Susceptible to human viral infection
2. *Continuous Cell lines*						
A 9	Fibroblast	Mouse sub-cutaneous	Adult	Neo	Anenploidy	HGPRT -ve; deriv L 929
BH K21 C 13	Fibroblast	Syrian hamster kidney	Newborn	Nor	Aneuploidy	Transformable by polyoma virus.
BR 13A	Epithelial	Rat liver	Newborn	Nor	Diploidy	Produce MSA
CHOK 1	Fibroblast	Chinese hamster ovary	Adult	Nor	Diploidy	Simple karyotype
EB 3	Lymphocytic	Human	Juvenile	Neo	Diploidy	EB virus +ve
GH1, GH3	Epithelial	Rat	Adult	Neo	Aneuploidy	Produce growth hormone
HeLa	Epithelial	Human	Adult	Neo	Aneuploidy	G6PD Type A
L 1210	Lymphocytic	Mouse	Adult	Neo	Aneuploidy	Rapidly growing suspension
L 5178Y	Lymphocytic	Mouse	Adult	Neo	Aneuploidy	
L 929	Fibroblast	Mouse	Adult	Nor	Aneuploidy	Clone of L-cell

(Contd.)

Table 3. Cell lines and cell strains in regular use and their origin (Contd.)

Name	Morphology	Origin	Age	Tissue	Ploidy	Characteristic
LS	Fibroblast	Mouse	Adult	Neo	Aneuploidy	Grow in suspension deriv. L 929
MCI⁻	Epithelial pleural effusion	Human breast	Adult	Neo	Aneuploidy	Estrogen recep. +ve
P 388D	Lymphocytic	Mouse	Adult	Neo	Aneuploidy	Grow in suspension
8180	Fibroblast	Mouse	Adult	Neo	Aneuploidy	Cancer chemotherapy, screening
3T31.1	Fibroblast	Mouse suriss	Embryonic	Nor	Aneuploidy	Adipose diff. [GREENAT at., 1978]
3T2A31	Fibroblast	Mouse BALB/c	Embryonic	Nor	Aneuploidy	Contact inhibited' readily transformed
NRR49F	Fibroblast	Rat kidney	Adult	Nor	Aneuploidy	Induction of suspension growth by transforming growth factors
Vero	Fibroblast	Monkey kidney	Adult	Nor	Aneuploidy	Viral substrate and assay.

Table 4. Advantages and disadvantages of finite and continuous cell lines

Features	Finite	Continuous
1. Ploidy	Diploid Euploid	Heteroploid Aneuploid
2. Transformation	Normal	Transformed
3. Tomourigenicity	Non-tumourigenic	No-tumourigenic
4. Anchorage dependance	Yes	No
5. Contact inhibition	Yes	No
6. Density limitation of growth	Yes	No
7. Mode of growth	Monolayer	No (or less so)
8. Maintenance	Cyclic	Monolayer of suspension
9. Serum requirement (in simple media)	High	Low
10. Cloning efficiency	Low	High
11. Markers	Maybe tissue specific	Chromosomal, enzymic

QUESTIONS

1. How are animal cell lines characterised? List the origin and characteristics of any one commonly used finite cell lines, and one continuous cell lines.

(BU Nov. 1997)

2. Give an account on characterization of animal cell line.

SHORT QUESTIONS

1. Use of markers
2. Maintenance of cell lines
3. Suspension culture

(BU Nov. 1999)

7

Cloning and Selection of Specific Cell Types

7.1 CLONING

Pure cell lines can be obtained by cloning from a heterogenous culture. The success of this technique is limited in animal cell culture is limited due to the poor cloning efficiencies of primary cultures. Successful cloning depends on the isolation of variants from continuous cell lines. These variants include biochemical mutants and cell strains with marker chromosomes.

(i) *Dilution cloning*: Dilution cloning method was developed by Puck and Marcus in 1955. The cells were trypsinized to produce a single cell suspension in this method. The trypsinization produce clumps of cells and over trypsinization reduces the viable status of a cell in cloning. Cloning of new cell lines from a good single cell suspension can be obtained by necessarily optimizing different length of trypsinisation and different recipe to give the optimum plating efficiency. The typsinised cells are seeded at low density in multi-well dishes, petridishes or plastic bottles and incubate and then propagated. The seeded petridishes are kept aside for a week untouched. The colonies formed are isolated, the medium is replaced and culture is continued for further week. When colonies are not formed with in the week or seed again in culture for 3 weeks if necessary (Fig. 7.1).

7.2 SOMATIC CELL FUSION AND HAT SELECTION

Hybrid cells are obtained by the fusion of different types of cells. This has been successfully achieved in animals. Basski et al; (1961) and Sorieul and Ephrussi (1961) have discovered that somatic cell would fuse in the presence of sendai virus (Fig. 7.2).

Several significant applications are associated with somatic hybrid cells such as (1) study of the control of gene expression and differentiation. (2) To study the problem of malignancy (3) gene mapping (4) viral replication and (5) production of hybridomas for production of antibodies.

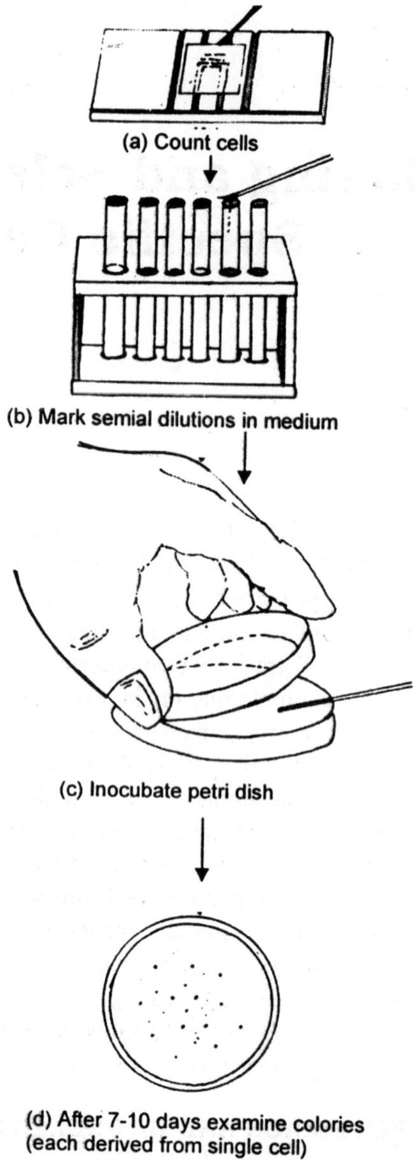

Fig. 7.1. Dilution technique for production of colonies from single cell.

There are examples of cell fusion within the body examples are multi nucleate muscle fibres are formed by the fusion of myoblast. They can undergo *in vitro* fusion to produce binucleate heterokaryones. Where there is fusion of cells but not nuclei or uninucleate hybrid cells.

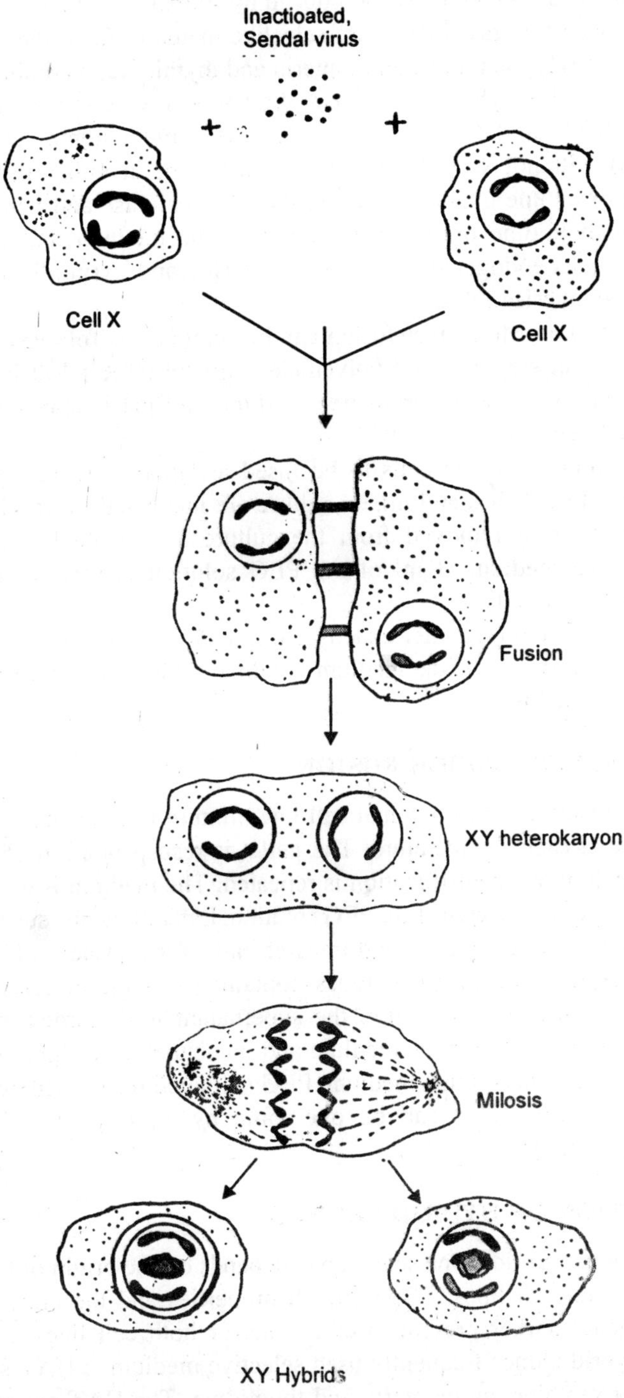

Fig. 7.2. Different steps for production of somatic cell hybrids using sendai virus.

The proportion of viable hybrids produced by somatic cell fusion are low. Due to this reason it is essential to use selective media to favor the survival of hybrids. HAT (Hypoxanthine aminopterin and thymidine) medium is used for selection of viable hybrids. Only heterokaryon's survive in this medium. The parental cells and fusion products of the same parental cell type (homokaryon) will not survive as they are deficient in either thymidine kinase or hypoxanthine guanine phosphoribosyl transferase and hence they cannot utilise thymidine or hypoxanthine from the medium and because aminopterin blocks endogenous synthesis of purine and pyrimidines. They are unable to synthesise DNA.

Monolayer fusion: Monolayer fusion can be achieved as follows:

(a) In the initial step sterlised polyethelene glycol (PEG) 100 is mixed with equal volume of serum free medium and pH is approximately adjusted between 7.6 to 7.9.

(b) Equal numbers of the cells to be fused are inoculated into 50 mm tissue culture dishes and mixed culture are incubated overnight.

(c) The medium is removed from the culture and washed once with serum free medium. 3 ml of the PEG solution is spread over the monolayer of cells.

(d) The PEG is removed after exactly 1 min and the monolayer is rinsed with 10 ml of serum free medium for three times before returning to complete medium.

7.3 MEDIUM SUSPENSION FUSION

In suspension fusion the two parental cell line mixture is centrifuged at 150 g for 5 minute at room temperature. The pellet is resuspended in 15 ml of serum free medium and centrifugation is repeated. The medium is discarded and the cells are resuspended in 1 ml PEG solution. Later dilute the suspension with 9 ml of serum free medium and transfer half of the suspension to two universal containers or centrifuge tubes containing 15 ml of serum free medium. Finally after centrifugation the supernatant is discarded and the pellet is resuspended in complete medium and incubated over night. Further these cells are cloned in solution medium. PEG mediated fusion will generate number of variations. This can be overcome by adding 10% DMSO. (Dimethylsulphoxide)

7.4 SELECTION OF HYBRID CLONES

(i) *HAT Selection:* Hybrid clone selection is depends on the origin of the two species of parental cell lines. Their growth properties and the presence of selectable genetic marker present in either one or both cell lines. For the selection of hybrid clones frequently used selective medium is HAT system consist of hypoxanthine, aminopterin and thymidine. The HAT system can be used to isolate hybrids made between mutant cell lines deficient in the enzymes like thymidine kinase, hypoxanthine guanosine phosphoribosyl transferase respectively.

In a classic example the cardian glycoside outbain, an antimetabolite can be exploited to select hybrids between rodent cells and cells of other species including human. Rodent cells are resistant to this antimetabolite up to 2 mM, while human cells are killed at 10^{-5} M outbain. On comparison hybrids exhibit more resistant to outbain than human cells. In another example HGPRT deficient rodents cell lines can be fused to unmarked human cells. These hybrids can be selected in the HAT medium contain low concentration of outbain.

7.5 PRODUCTION OF MONOCLONAL ANTIBODIES (Hybridoma Technology)

The hybridoma is a hybrid cell (heterokaryons) is a result of fusion of myeloma cells and immuno competent spleen cells. The resulting hybrids are selected in HAT medium and subsequently screened for the production of a specific antibody. The hybrid cells thus produced possess the ability to produce antibodies due to the β-lymphocyte genome and indefinite growth *in vito* due to tumor (myeloma) cells involved in the fusion. Thus the fusion produces are capable of synthesizing immunoglobulin, these specific single antibodies are called "monoclonal antibodies".

The steps involved in this technique are outlined as follows:

(1) A mouse is first-immunised with crude antigen and later the spleen is removed.

(2) It is then minced and placed in culture with TK^- mouse myeloma cells in the presence of polyethylene glycol. The cells get fuse, and fusion products are cloned in HAT medium.

(3) The parental cells or homokaryons will not grow in HAT medium because the myeloma is TK^- and the spleen cells are unable to grow *in vitro*.

(4) The clones are then grown and their specificity tested by immunoassay with their suprenatant medium using appropriate target cells or antigen, anchored to microtitration plates, where a specific antigen-antibody complex forms. It can be detected autoradiographically with ^{131}I-labelled or enzyme-conjugated antimouse immunoglobulin.

(5) The required clones are recloned, retested and then frozen. For subsequent antibody production, the cells may be propagated in suspension culture and the supernatant medium used or they may be passaged in mice as an ascites tumor, ascitic fluid gives very high yields of antibody although contaminated by globulins from the host animal.

The hybridoma technology is a potential technique for the study of immuno system, production of cell type – specific markers, performance of structural analysis of proteins, in the production of vaccines and in purification of protein by immunoaffinity chromatography.

The following protocol for preparation of hybridomas have developed by chrismorris, European collection of animal cell cultures, PHLS, CAMR, Porton own, Salisbury, England.

(1) Culture myelomas in HT medium and maintain between 2-6 × 10^5 cells/ml for 7-10 days prior to a fusion.

(2) Inject 5 ml of ice cold 0.34 M sucrose solution to healthy mice (4-8 week old) and aseptically remove peritoneal macrophage cells. Count the cells and seed at 10^5/well.

(3) After the final antigen injection, successfully immunized animals are sacrificed and aseptically remove the spleen and transfer to growth medium. Tease spleen with surgical needles, transfer the dispersed cells into medium.

(4) Mix myeloma and spleen cells in appropriate ratio (eg. 5 × 10^7 myelomas to 10^8 spleen cells). Centrifuge and resuspend both myeloma and spleen cell pellets in 20-30 ml of both serum free medium in 50 ml tube.

(5) Hold the tube in water at 37°C and add dropwise 1 ml of PEG solution onto the wall of the tube just above the pellet over a 5-10 days period.

(6) Decant the medium containing PEG. Resuspend the cells by gently adding HAT medium.

(7) The colonies can be observed after 3-4 days. Maintain expanding cultures in HAT for at least 10-14 days and change medium (50%) growing wells at least once prior to screening.

QUESTIONS

1. "Technique of cell fusion and use of HAT selection system helped in production of monoclonal antibodies" – Discuss.
2. Explain the production of monoclonal antibody by hybridoma technique. (BU Nov. 1999)
3. What are selection markers? Explain their relevance illustrating at least two specific examples. (BU April 2000)

SHORT QUESTIONS

1. HAT selection
2. Application of cell fusion (BU May 1996)
3. Cell fusion (BU Nov. 1999)

8

Organ Culture

A pieces of organ can be cultured on the artificial medium *in vitro*. The objective of culturing organs is to maintain the architecture of the tissue and direct it towards normal development like *in vivo*. It is necessary that the tissue should not be disrupted or damaged during organ culture, so care has to be taken to achieve this. Unlike tissue culture, the organ culture requires a more careful manipulation than tissue culture technique discussed in the earlier chapters. The media for organ culture would be same as described for cell and tissue culture except few changes or addition of few ingredients. It is easy to culture embryonic organ than the adult animals. The adult organ requires a high amount of O_2 (about 95%) special apparatus and serum free media (T8) are required for the adult organ culture.

8.1 CULTURE OF EMBRYONIC ORGANS

The embryonic organs can be cultured using following three techniques.

8.1.1 Organ culture on plasma clots

A plasma clots can be prepared by mixing 15 drops of plasma with five drops of embryo extract in a watch glass on a moist cotton wool to prevent excessive evaporation from the dish on the top of the plasma place a small piece of organ tissue. Now this technique has been modified. The raft of lens paper or ryon net is used on which the tissue is placed. The raft makes the easy movement of the tissue and also facilitates feeding of the culture, so that excess fluid can also be removed.

8.1.2 Organ Culture on Agar media

The solidified nutrient agar media, which may or may not contain serum can be used for organ culture. The medium with agar has the advantage that the support for organ culture, it does not liquify and no additional mechanical support is required, usually adult tumor cells fail to survive on agar media, whereas embryonic organs grow well. The constituents of nutrient agar medium consist of ingredients of agar (1% in basal salt solution), chick embryo extract and horse serum in the ratio of 7:3:3.

8.1.3 Organ culture in Liquid media

The liquid media consist of all other ingredients except the agar. In this method, generally perforated metal gauze or raft made of lens paper or cellulose acetate net is used. These may provides support to the organ.

8.2 WHOLE EMBRYO CULTURE

Spratt (1956, 1957), studied, how metabolic inhibitors affect the development of embryo *in vitro*. In this technique 40 hours old embryos were used and the embryo development could be followed for another 24-48 hours *in vitro* before the embryo dies. For embryo culture a suitable medium prepared is poured into a watch glasses, which are then placed on moist absorbent cotton wool pad in petri dishes. The eggs are incubated at 38°C for 40-42 hrs, so that about or dozen embryos could be produced. The egg shell is sterilized with 70 per cent ethanol and broken into pieces and transferred into 50 ml of BSS. The vitelline membrane covering the blastoderm is removed and kept in petri dish containing BSS. The adherent vittellin membrane is removed with the help of forceps and the embryo is examined under the microscope to determine the stage of development. The blastoderm is then placed on the medium in the watch glass placed on sterile absorbent cotton wool pad in petri dishes. Excess of BSS is removed from medium and incubated at 37.5°C for further development.

8.3 CULTURE OF ADULT ORGANS

8.3.1 Culture of adult liver

Berry and Friend (1969) studied with collagenase and hyaluronidase, these provides a technique whereby or good viable suspension of liver parenchyma cells could be plated out with high purify and could get or viable monolayer and also epithelial cultures may be propagated by this technique.

Proteolytic Enzymes such as collagenase, which are relatively noncytotoxic, could be used to disrupt intercellular junction and liver. These enzymes digest the connective framework with in 15 min. These enzymes to work properly, the liver should be free from blood and depleted and ca^{2+} by washing with calcium free buffer. After the enzyme treatment collect the disrupted cells by three differential centrifugation and wash the cells and court the viable hepatocytes. Collect these hepatocytes in Ham's F_{12} or Williams' E medium enriched with 0.2% bovine albumin (grade V sigma) and 10 μg/ml bovine insulin.

8.3.2 Culture of Pancreas Cells

Wallace and Hegree (1979) first time produced epithelial monolayer from fetal rat pancreas by primary explant technique using collagenase. These cultures remained free of fibroblast for several days, and contained many endocrine cells.

The pancreatic acinar epithelia are fractionated and inoculated onto collagen coated culture dishes. The fractionated cells were filtered through cheese cloth or nylon sieves before inoculation onto collagen coated culture dishes. Layer the suspension over a BSA solution. Inoculation densities could be $3 \times 10^5/cm^2$. The cell will adhere as aggregate colonies within 72 hr.

8.3.3 Culture of Kidney Cells

Structurally kidney is the complex organ in which the system of nephrons and collecting ducts are made up of numerous functionally and phenotypically distinct segments. This segmental heterogeneity may due to cellular diversity, still this has to be fully characterized. Some of the tubular segments possess several morphologically distinct cell types. This changes in cells presents a challenge to the scientist in isolating pure or highly enriched cell populations.

Primary culture of tissue fragments are excised from the outer cortex and human kidney. These tissue fragments are trasferred to 50 ml tube containing 20 ml worm collagenase—typsin solution and incubate at 37°C incubator in orbital shaker. The tissue is periodically pipetted and collected the dissociated cells by centrifugation. The cell suspension is filtered through size limiting screen to remove undigested fragments. The cells are then washed to remove the enzyme, suspended in serum-supplemented medium and seeded on the culture medium.

QUESTION

1. Describe various techniques involved in organ culture.

SHORT QUESTIONS

1. Plasma clot
2. Culture of kidney cells

9

The Transformed Phenotype of Animal Cells

'Transformation' in microbiology means, the uptake of new genetic material. This process in mammalian cells is called as "transfection".

In animal tissue culture, transformation means spontaneous or induced permanent phenotypic change that does not necessarily involve the uptake of new genetic material. The transformation results due to several reasons. (1) infection with a transforming virus such as polyoma, (2) from incorporation of new genomic DNA, (3) it can also arise spontaneously or following exposure to a chemical carcinogen. Recent studies have shown that some of the phenotypic properties of transformed cells can be restored to normal by chemical or hormonal inducers.

Transformation results in three major classes of phenotypic changes, one or all of which can be expressed in one cell strain. They are:

(i) *Immortalization*; is acquiring of an infinite life span. This is due to: (1) the deletion or mutation of one or more senescence genes (2) over-expression or mutation of one or more oncogenes and immortal cell line fusion will generate finite span hybrids suggesting that the senescence genes are dominant.

Immortalisation, *per se* (in itself) does not imply either failure in growth control or the development of malignancy.

(ii) *Aberrant growth control*: This may arise simultaneously with immortalization, or also occur as a distinct subsequent event. Cell lines such as the 3T3 group are immortal but still maintain contact inhibition of cell motility and density limitation of cell proliferation. Immortalised cell lines can subsequently lose their growth control, and (1) grow to a higher saturation densities (2) clone in agar, and (3) grow on confluent monolayers of homologous cell. These cells requires lower serum or growth factor. They acquire autonomous growth control by over expression of oncogenes or by deletion of suppressor genes.They can also form clones with a higher efficiencies.

(3) *Malignancy:* It is stage reached by the cells that have developed the capacity to generate tumors, if implanted *in vivo* into an isologous host or transplanted as a xenograft into an immune deprived animal.

Normal human urothelium made up of predominantly diploid cells, that are nonangiogenic, non-envasive, and nontumorigenic. Here *in vitro* transformation is a progressive, multi step event, over a period of up to 2 years, these lines becomes first angiogenic, then aneuploid and ultimately tumorigenic and invasive.

Hence, transformation is apparently a multi step process often culminating in the production of neoplastic cells. It is therefore find that cell lines from malignant tumors, already transformed can undergo reduced anchorage dependence, more pronounced aneuploidy and immortalization. This type of transformation was observed in a squamous lung carcinoma cell line in IAN FRESHNEY Laboratory, and the resultant continuous cell line actually lost tumorigenicity is nude mice.

9.1 GROWTH ON CONFLUENT MONOLAYER

Aaronson et al; (1970) showed that transformed mouse 3T3 cells and human fibrosarcoma cells were able to form colonies on confluent monolayer of normal cells, which normal fibroblast were not. Cells cultured from both normal and malignant breast tissue formed colonies on contact inhibited monolayer of human fetal intestinal cells (FHI) as do normal epidermal cells. Rheinwald and Green (1975) had previously shown that normal epidermal cells grew on 3T3 monolayers. This technique may discriminate between malignant and normal cells, but this requires that the normal cells are of the same lineage as the malignant cells.

In case of human glioma will grow on confluent glial feeder layer, but cells derived from normal glia will not grow, while some normal cell lines will grow on other feeder layer eg. Fetal lung fibroblasts and FHI.

Preparation outline is as follows (protocol):

(1) Seed flasks with feeder cell (3T3 or other cell line that will become contact exhibited after reaching confluences)

(2) Feed cultures, until all available substrate is covered with cells and mitosis is greatly reduced, treat cells during exponential growth with mitomycinc.

(3) Trypsonize putative tumor cells and suitable control cells and dilute to 10^4, 10^3 and 10^2 cells per ml in 20 ml and replaces medium in each pair of feeder layer flasks with 5 ml of cell suspension.

(4) Incubated for 2-3 week with medium changes

9.2 TUMORIGENESIS

The definition of malignancy is the formation of invasive or metastasizing tumors *in vivo*. Transplantable tumor cells (10^6 or less) injected into isogeneic hosts will produce invasive tumors in high proportion of cases. To study the tumorigenicity of human tumors, a number of models have been developed using immune suppressed or immune-deficient host animals. Some tumor cell lines and tumor biopsies have failed to produce tumors as xenografts and frequently fail to metastasize, although they may be invasive locally.

QUESTIONS

1. What are the important phenotypic changes associated with transformed cells?
2. What is transformed cell line? Explain the origin and characteristics of any two transformed cell line.

(BU Nov. 1999)

SHORT QUESTIONS

1. Transformed cells

(BU April 2000)

2. Growth on monolayer
3. Tumerogenesis

10

Transfection of Animal Cells

One of the most important applications of animal cell and tissue culture is the transfer of foreign genes into our livestock. In organisms like bacteria and other microbes and even in higher plants, the uptake of genes by cells is often described by the term transformation. However in animals this term has been called as transfection, because the term transformation in animal cell culture is used to describe phenotypic alteration of cells.

Transfection or gene transfer in animals may be carried out at the cellular level to get transfected cells, which may be used for a variety of purposes such as production of fine chemicals and pharmaceutical drugs. It can also be undertaken for basic studies involving study of structure and function of genes. Many mammalian cell lines have been regularly utilised for these purposes. Transfection has also been achieved successfully for the production of transgenic animals.

A gene of interest can be cloned and transformed into host cell to investigate functions of individual gene. Variety of techniques such as transfection, lipofection, retroviral infection have been in use. Cloned DNA is often conveniently maintained as part of a bacterial plasmid. Many plasmid can obtain high copy number during bacterial growth, thus ensuring a plentiful stock of DNA for experimentation. Plasmid DNA is purified from the bacteria prior to use. Once the sequence of interest has been cloned, it can be manipulated further to isolate subclones containing promoter sequence. By genetic manipulation promoter sequence can be linked to a reporter gene (Eg. β-gal) whose product (β-galactosidase) can be readily assayed subsequent to transfection. In this way, tissue specific gene expression can be analysed in detail. Oncogenes and tumor suppressor gene function can also be analysed by similar manipulations.

The improvement in livestock through transgenesis already led to the following encouraging results

(1) increased milk production in cattle

(2) increased growth rate in fish
(3) production of large scale valuable proteins in milk, urine and blood, of livestock, enabling the use of transgenic animals as bioreactors for molecular farming.
(4) improvement of wool production in transgenic sheep. These encouraging results have made transfection and production of transgenic animals or facinating thrust area of research.

10.1 GENE TRANSFER OR TRANSFECTION

In the earlier chapter, we have discussed about the technique of somatic cell fusion as a means of transfer of the whole genome from one organism to another. However, fusion can also be obtained between isolated nuclei, or between a nucleus and an enucleated whole cell. The transfer of genetic material to whole cells can be exercised at the level of (1) individual chromosomes or fragments (or) (2) isolated genes or gene fragments

10.1.1 Transfection of Fertilised Egg or Embryos

The production of transgenic animals requires transfection of specialised cells or embryos, since only egg or embryos can develop into whole animals. The egg may be modified by transfer of whole nuclei, whole chromosomes or parts thereof or DNA fragments.

10.1.1.1 *Transfer of whole nuclei or Split embryos*: Transfer of whole nucleus from a somatic cell of the superior donor to the enucleated egg can be achieved using the following steps (Fig. 10.1).

(1) Centrifuge cytochalasin-B treated cells to make enucleation of unfertilised egg. After the centrifugation the nuclei will detach from the eggs and pellet at the bottom of the tube, leaving enucleated eggs in the supernatent.
(2) Karyoplasts (nuclei with only some residual plasma membrane) are similarly obtained from the blastula stage of the developing embryos of the donor.
(3) Karyoplasts from the donor are incubated with enucleated eggs in the presence of PEG to achieve fusion.
(4) The manipulated egg is transferred to the uterus of a surrogate mother for development.

This technique can be employed, to where a superior developing embryo may be bisected into two parts using surgical blade. Each bisected part of the embryo may be separately transferred to an enucleated unfertilised egg (Fig.10.2). Manipulated egg may then transferred to the uterus of a surrogate mother for further development.

10.1.1.2 *Transfer of whole individual chromosomes or fragments*: Chromosomes may be isolated from metaphase cells by hypotonic lysis. Incubate isolated chromosomes with whole cells after coprecipitation with calcium phosphate results, in their incorporation into the nuclei. The chromosome once isolated may also be subjected to fractionisation using density centrifugation or flow cytophotometry and individual chromosome pairs may be inserted into recipient cells.

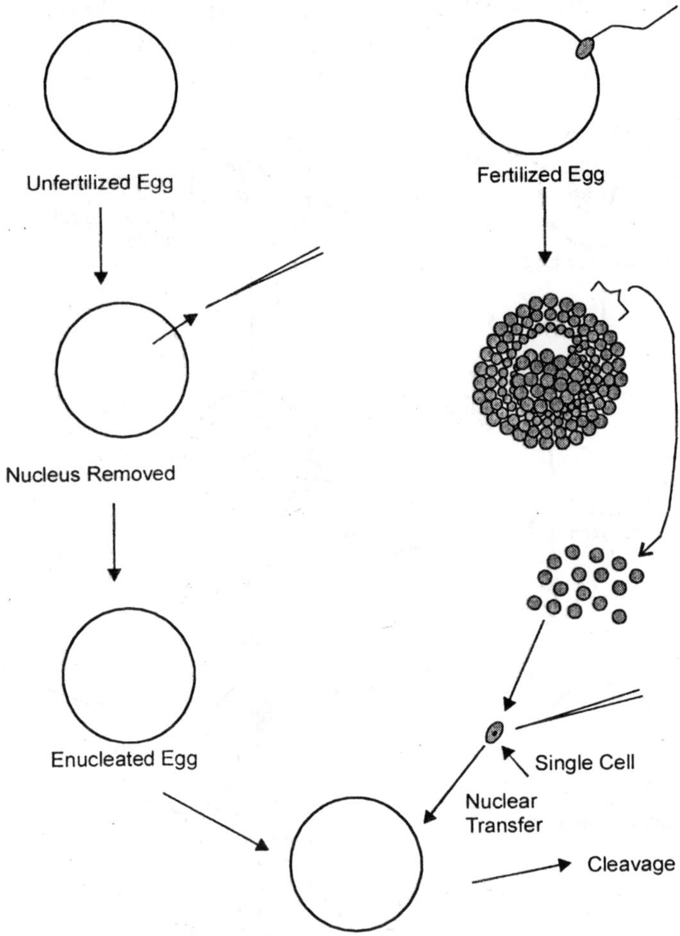

Fig. 10.1. Fusion of karyoplasts with enucleated cells in the presence of PEG.

10.1.1.3 *Microinjection of DNA into nucleus and egg*: In this method, DNA is directly injected into the nucleus of a cell or into a fertilised egg of one to two cell. The microinjection assemble consists of a low power sterioscopic dissecting microscope, to view the ovum and the entire process and two micromanipulators, one for a glass micropipette to hold the ovum by partial suction and the other for a glass injection needle to introduce the DNA into the male pronucleus. The male pronucleus is much larger than the female pronucleus of fertilised mammalian ova.

The general procedure for microinjection is explained in the figure 10.3. Superovulation was induced to the donor females using appropriate hormone treatment. PMSG (pregnant maare serum gonadotrophin) which stimulates growth and development of follicles which contain the developing oocytes were given to female mice. HCG was given to release the oocytes. The superovulated females are then mated with fertile male and large numbers of fertilised one to two cell ova/embryos are collected surgically.

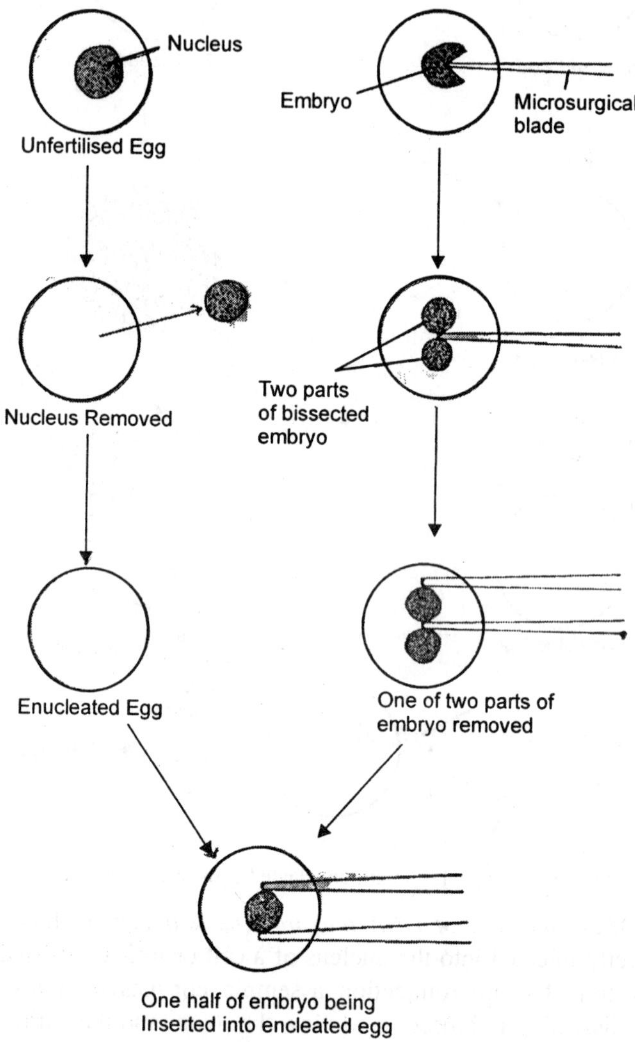

Fig. 10.2. Splitting of an embryo, followed by transfer of one half of the embryo to an unfertilized egg whose contents were removed.

The DNA is microinjected into a fertilised egg before the fusion of male and female nuclei. The egg is first immobolised by applying mild suction to the large, blunt holding pipette and DNA is then injected through the sharp end of a narrow glass micro needle. In mice, the microinjected embryos are cultures *in vitro* upto the morula or blastocyst stage. The surviving embryos are then transferred into the uterus of surrogate mother, these embryos develop to full term and give rise to normal mice. A proportion of the progeny so produced will be transgenic in that all their cell will contain the transgene stably integrated into their genomes. But in case of fish, the microinjected embryos are incubated in water until hatching.

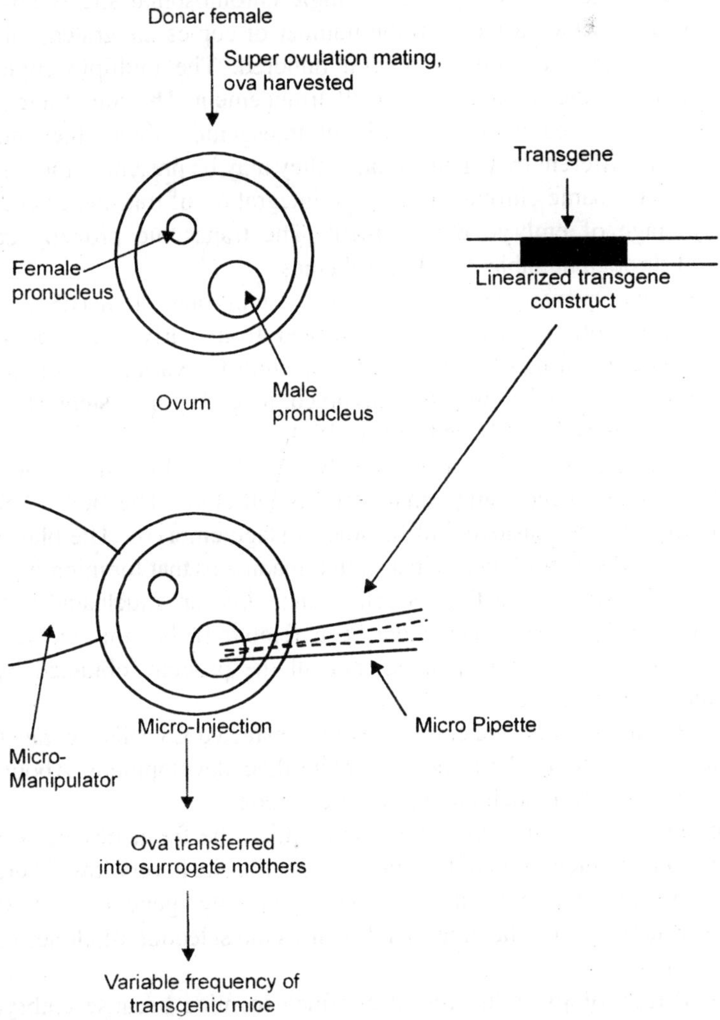

Fig. 10.3. Microinjection technique for transfection to produce transgenic mice.

The transgene construct is prepared in a buffer solution and is injected into egg. Typically 2pl (picolitre) of the DNA solution is injected in a pronucleus. But in case of fish, 20 nl (nanolitre) DNA solution, containing 10^6-10^8 linearized transgene constructs, is injected into the cytoplasm of a single ovum.

An average of about 3-6% of the progeny derived from microinjected embryos are transgenic, the frequency is much less in other animals eg. < 1% in sheep and pigs. In fish about 35-80% of the embryos survive of which 10-70% may be transgenic. The transgenes in transgenic animals show typical mendelian inheritance.

The transgene integration occurs at random sites in the genome, but in a given cell or embryo usually only a single chromosomal site is involved. There will be a wide variation in the number of copies integrated, this may be ranging from one copy to several hundred. The multiple copies are integrated at a single site in a head to tail arrangement. The site of integration of transgene may be varied in different transgenic animal, they may be present in the different transgenic animal, they may be present in the different locations of the same chromosomes. The integration of transgenes occurs at an early stage of embryo development. The transgenic progeny can be recovered through suitable breeding schemes.

10.1.1.4 *Embryonic Stem Cell transfer:* Embryonic stem (ES) cells of mice are pluripotent cultured cells derived from early pre-implantation embryos. These cell can be preserved long time for various manipulations for gene transfer. Different steps involved in embryonic stem (ES) cell transfer technology is explained in Fig. 10.4.

The injection and coculture are the two methods, by which transgenic animal can be produced using transfected ES cell clone. The cloned ES cells are injected into the blastocoel of blostocyst stage embryos. The blastocysts are obtained from donor females in a similar manner as that for microinjection. It is possible to transplant the ES cell nucleus into an enucleated fertilised ovum (an ovum whose nuclei have been removed), but this technique is quite tedious and it is not adopted, but all the progeny obtained by this technique are transgenic.

The cocultured or injected embryos with transfected ES cells are transferred into surrogate mother where they complete their development. About 30% progeny derived from such embryos are chimeric.

Since DNA integration in genome is random, the frequency of targetted gene transfer i.e. integration of transgene at specific site is very low. Therefore, the microinjection will not suitable to identify targetted gene transfer. But ES cell line technology, on the other hand permits the selection of clones having targetted gene.

ES cell technology is at present confined to mice because embryos of most other mammals do not survive *in vitro*, so that it is extremely difficult to produce and maintain their ES cell lines. Recently ES cell line have been developed in some mammalian species other than mice, but no chemeric individual have not been produced using them so far. In future it may applicable to mammalian species.

10.1.1.5 *Lipid-mediated gene transfer* (lipofection): The use of cationic lipids for DNA transfection into mammalian cell has become widespread, because they are safer than viral vectors, can be produced in large quantities, and can deliver large DNA fragments of up to several megabase pairs long into cells.

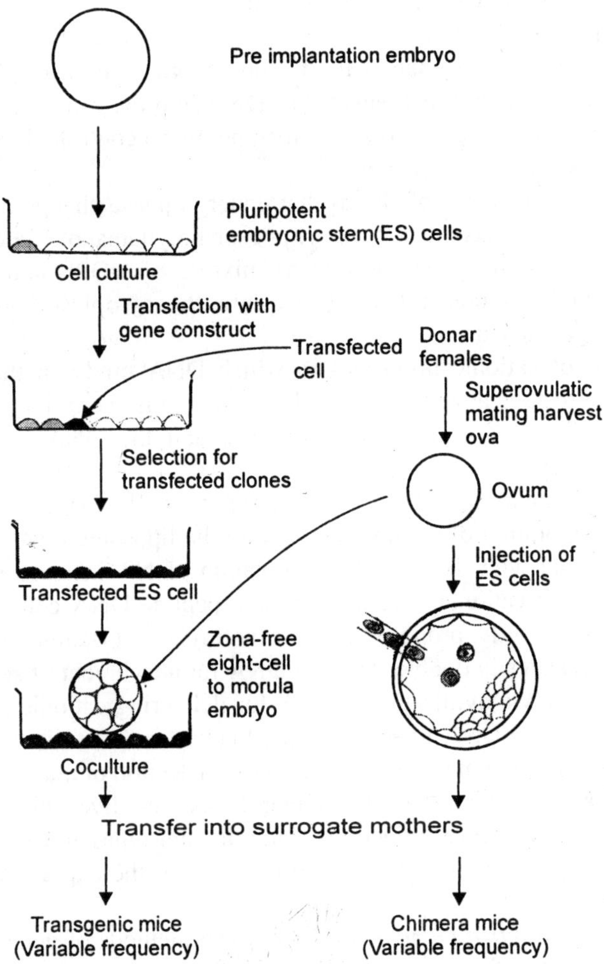

Fig. 10.4. Embryonic stem (ES) cell transfer technology

The delivery of DNA into cells using *liposomes* is called lipofection. Liposomes are small vehicles prepared from a suitable lipid. The use of cationic lipids for the construction of liposomes is a distinct advantage as DNA spontaneously and efficiently complexes with these liposomes. The cationic liposomes have a single lipid bilayer membrane (unilamellar), and they bind to the cell efficiently. Probably they fuse with the plasma membrane and there by deliver the DNA (complexed with them) into the cells, which brings about transfection. Following procedure is followed for preparing liposomes for delivery of DNA.

(1) Sonication will be used to disperse the phospholipid like phosphatidyl choline (PC). This will destroy DNA. DNA upto 1 kb can be incorporated into small sonicated liposomes. Exposure of the anionic lipid phosphotidyl serine (PS) to Ca^{2+} will allow the entrapment of large DNA sequences into liposomes.

(2) The fusion proteins of sendai virus is incorporated into the liposome membrane; this enables the fusion of liposomes with plasma lemma and thereby a direct delivery of DNA into the cytoplasm. A receptor protein is also incorporated which permits a controlled delivery of the DNA into the target cells.

(3) The use of ionizable lipids that undergo phase change in response to pH of cytoplasm e.g. dioleoyl phosphatidyllethanolamine (DOPE), to construct liposomes usually as mixture of DOPE and PC is used. Such liposomes release DNA into the cytoplasm once they are phagocytosed.

(4) Use of cationic liposomes to which DNA binds on the outside by electrostatic attraction. These liposomes cause perturbations in plasma membrane due to which they fuse and the DNA enters into the cytoplasm.

(5) The liposomes may be targetted to cell surface receptors by incorporation of ligand proteins into the liposome membrane. This is possible with the liposomes containing DNA inside the vesicle but not with cationic lipid and cationic peptide DNA complexes. It has been recently proposed to use biotinylated bisanthracycline which intercalates in double stranded DNA for attachment of specific ligand proteins via avidin to the DNA for delivering cationic lipid/peptide-complexed DNA to specific target cells.

Lipofection is the method of choice for transfection of mammalian cells *in vitro*. It has also been used to deliver DNA into live animals by direct injection or intravenous injection. Cationic liposomes have been used in extravenous or intratracheal injection in mice for the expression of marker genes in lungs.

10.1.1.6 *Targetted Gene transfer:* In most methods of gene transfer, the integration of the foreign gene occurs at apparently random chromosomal positions. There misplaced genes can be subject to correct regulation under the control of regulatory sequences.

The gene transfer at homologous sites in the host genome is popularly known as targeted gene transfer. This is achieved by a process of homologous recombination by which the transgene replaces its allele in the genome (Fig. 10.5). Homologous recombination may be defined as the exchange of identical segments between two DNA molecules that have identical or almost identical sequence. The frequency of integration by homologous recombination appears to be only 0.1 to 1% of random integration events;

this makes the recovery of transgenics representing targetted gene transfer quite difficult.

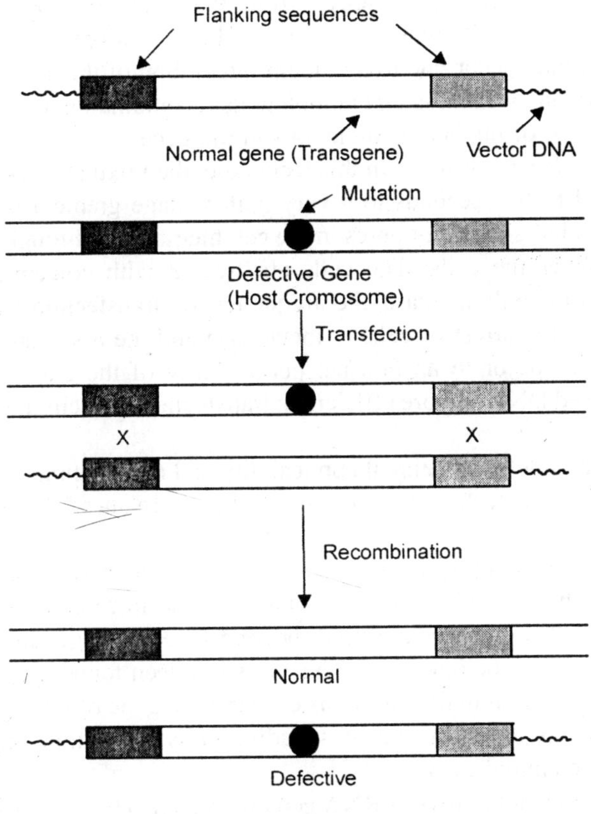

Fig. 10.5. A Schematic representation of targetted gene transfer through homologous recombination.

Oliver Smithies in 1985 at Madison (USA) was successfully achieved targetted gene transfer in human cells. He has transformed a human β-globin gene into the β globin gene of recipient cells by recombination. The targeted gene transfer was possible due to the presence of homologous DNA sequences at the targeted site and in the vector carrying the foreign gene. The marker genes are exploited for selection of cells, they include (1) by assaying the activity of an enzyme, as in hypoxanthine phosphoribosyl transferase (HPRT) enzyme (2) use of antibiotic resistance genes (3) by polymerase chain reaction (PCR). If sequence of the gene is available for making PCR primers (4) Chimeric mice have actually been prepared by targeted gene transfer where HPRT gene or homoeobix containing genes have been modified in some embryonic cells. If these cells contribute to the production of egg or sperm, the transgenic mice will be produced.

Gene replacement or targetted gene transfer offers two unique advantages. First, it places the functional gene into the same context i.e. the transgene is flanked by the same DNA sequences, as the defective mutant gene it has replaced. It is well recognised that the sequences flanking a gene. Even those located at considerable distances from the gene, usually influence its expression. Second, it precludes the possibility of disruption of an important gene which may occur due to a random integration of the gene. However, it suffers from one weakness that homologous recombination usually promotes the occurrence of unplanned mutations in the gene.

10.1.1.7 *Electroporation*: In this technique, the mixture of cells and DNA are exposed to few seconds to a very high voltage gradient (4000-8000 v/ cm). This induces transient pores in the cell membranes through which DNA seems to enter the cells. The cells are treated with colcemid before the electroporation will increase the frequency of transfection. This is most likely due to the arrest of cells at metaphase and the associated absence of nuclear envelope or to an unusual permeability of the plasma membrane. The linearised DNA is more efficient in transfection than circular supercoiled DNA.

This technique has a general applicability and many animal cell types that could not be transfected by other approaches were successfully transfected by this approach.

10.1.1.8 *Retroviral infection of embryo*: Recombinant retroviruses produce virions which are used to infect animal cells and mice embryos. Generally early 4-16 celled embryos are used; beyond 4 celled stage, an embryo may not be injected by the retrovirus. This virus has been found to be an efficient vector system for animals. The virus carrying the gene of interest transfers it into the genome of embryonic cells leading to its integration and production of transgenic animals.

The recombinant retrovirus RNA genome is copied by reverse transcriptase to yield a DNA copy which becomes integrated into the cells genome. The reverse transcriptase is encoded by retrovirus and is produced immediately after injection. However, the reverse transcription can occur only in those cells which go through S phase i.e. are mitotically active. Recombinant retrovirus integrates its DNA into the cellular genome at random sites and usually is not accompanied with deletions or rearrangements.

10.1.1.9 *Calcium phosphate precipitation*: In this method, the DNA to be used for transfection is first dissolved in a phosphate buffer. Calcium chloride solution is then added to the DNA solution. This leads to the formation of insoluble calcium phosphate which co-precipitates with DNA. The calcium phosphate DNA precipitate is mixed with the cells to be transfected. The precipitate particles are taken in by the cells by phagocytosis. On a small portion of the transfected cells, the DNA becomes integrated into cell genome producing stable or permanent transfection. This procedure has been modified to obtain transfection from 2-20%. The use of this method is limited, this can be applied to mammalian cells; but many cell lines fail to adopt to this method.

QUESTIONS

1. What is transfection? Give an account on types of transfection in mammalian system.
2. Describe different gene transfer system for mammalian system.

11

Tissue Engineering

Tissue engineering (TE) is the application of engineering principles to cell culture for the purpose of constructing functional anatomical units, normally of constructing for reconstructive surgery. There are 3 main components are involved here (1) cell biology, (2) material science (polymer/protein biochemistry) and (3) biological engineering.

Tissue engineering aims to supply body parts for repair of damaged tissues and organs, without causing an immune response or infections or using cadaveric tissue or mutilating other parts of the patient. Tissue engineering leads to the biological integration of tissue implants of surgery, with ease, stability and safety of patient.

In this technique living cells possess an innate and self-sufficient cell types to a suitable support matrix which allows proliferation and give rise to an organised and functional tissue, resembling the tissue of origin. This is likely to be the simplest, most economic approach, where it can be applied.

Based on the understanding about tissue and regeneration biology it seems that epithelial and endothelial cell layer are proved better at organising themselves through innate cell behaviours. Cells such as keratinocytes and vascular endothelial cells make strong cell-cell interactions and form coherent sheets. On the other hand, connective tissues or mesenchymal layer, in which extensive deposition of collagen matrix, will not form an appropriate structures spontaneously.

The contribution of tissue repair processes is important in TE developments, these are almost all surgical implants, including skin, blood vessels, heart valves, nerves, joint surfaces and ligaments.

11.1 CELL SOURCES

The first requirement for cell culture in any TE applications is to generate sufficient cells to seed. It very critical that, the stage at which cells are to be introduces into the implant, required implant substrate, the degree of function and the timing are very important.

It is very important that, what kind of source of cells should be used, whether it is autologous, allogeneic or xenogeneic. The use of the patients cells (autologous transplant) is often the most straight forward option. A biopsy of the appropriate tissue is taken from the patient, enzymatically digested or explant cultured and the required number cells are taken. This method will avoid the immune response and the possible transfer of infection inherent in the use of allogeneic cells. It's drawbacks are include disease state and age of patient, it may not be possible to obtain adequate biopsy material.

Some cell types may pose problems, for example articular cartilage, which has a relatively low density of cells with a low mitotic capacity. Choice of skin for tissue engineering is very important criteria, the implanted cells should resume their original phenotype and continue to produce a thickened epidermis.

Allogeneic sources of cells have advantages over autologous cells. It can be relatively easy to obtain healthy donor tissue. The cells may be cultured on a large scale at central factory sites, with none of the time constraints of autologous cells. The product will be cheaper with a consistent quality and available as and when required by the patient. The major problem associated with allogeneic cells is graft rejection.

The level of immune response generates by cells differs between cell layers. For example, endothelial cells can induce a large reactions. Fibroblasts, keratinocytes and smooth muscle cells are less immunogenic. The reaction may also be dependent upon the donor age of the cells used, as fetal or neonatal cell can elicit little or no immuno response. There are many reports of reducing the immunogenicity of transplanted cells, with the limited success.

Gene therapy has a role to play in tissue engineering. Cell population can be manipulated by adding genes either to increase their output of a certain protein or add to their stock of expressed genes. Examples are alteration of keratinocytes to produce transglutaminas-1, which is deficient in patients with lamellar icthyosis, disfiguring dermal disorder. The engineered cell were transplanted in athymic mice and produced as normal epidermis. The technique has also been used to induce fibroblasts and keratinocytes to release factors they may not normally produce.

The culture of cells for TE implants will depend on the function of the cells. For many cell types, the classical monolayer may allow proliferation of cells, but also increases the possibility of phenotypic changes. Many cells revert to a fibroblast like morphology after long term monolayer culture (eg. chondrocytes) or they may loose their proliferative capacity (e.g. endothelial cells). This may cause the cells to loose their function due to the inability to produce certain hormones.

Culture medium is also important. Some cell types, for eg. chondrocytes survive at nutrient level, which would be critical to other cell types. Fibroblasts or chondrocytes cells grow in low density 3D aqueous gels. To promote cell proliferation growth factors and cytokinins can be added. For example mitogenic growth factors can be added to promote fibroblast proliferation.

11.2 TISSUE ENGINEERED SKIN

Tissue engineering of skin was first demonstrated in the year 1975. TE of skin was possible, when the sheets of human keratinocytes could be grown in the laboratory in a suitable form for grafting. It was demonstrated that when an irradiated 3T3 cells stimulated growth of skin cells and allowed their differentiation into epidermis. This technique has been extensively modified and applied clinically, but as a skin treatment without a dermal layer. 3T3 cell are believed to secrete factors which help in growth and differentiation. This technique was used by BTI (Biosurface Technology), where postage stamp sized skin biopsies are taken and expanded 10,000 times within three to four weeks. This gives only the cultures epidermis, which when applied to burnt area, helps regeneration of dermis underneath.

In case of severely burned patients, skin biopsies may not be available from the patients, following two approaches can be used to overcome the problem. (1) Skin from cadavers (dead bodies) are taken to cover and protect against infection and desiccation, until their own skin is grown (2) Neonatal foreskin (loose skin from the tip of penis of newly born child) is cultured which creates artificial dermis layer. This leads to allogenic grafts stimulating rapid healing. New born skin cells are used, since they grow relatively more vigorously. The use of synthetic polymers by ATS company will allows the cells to grow into skin without scars, so that it is believed that this technology will be basis for future organ development.

The explant for preparing artificial skin for the graft must come from the patient himself to avoid graft rejection. 3 cm^3 skin explant can yield about 1.7 m^2 artificial skin in 3-4 weeks representing a 5000 fold increase. In about 5 years after grafting of the artificial skin, all the essential components of skin are regenerated. Artificial skin grafts have been used to successfully repair several types of skin defects, including chronic skin ulcers.

A variety of diseases have also been treated using cultured keratinocytes. Following are some examples.

(1) Scars on skin can be removed using cultured skin.

(2) Cultured oral keratinocytes have been used to regenerate epithelium of mouth.

(3) Cultured urethral keratinocytes have been used to repair congenital penile defects.

(4) For some middle ear diseases with troublesome discharge also, cultured epidermal keratinocytes are used.

(5) Chronic skin ulcers have also been treated with cultured grafts; Even allografts (skin from another individual) are successful in these ulcers.

11.3 TISSUE ENGINEERED UROTHELIUM

Human urothelial cells and bladder smooth muscle cells can be cultured *in vitro*. It is likely that construction of tissue engineered urothelial implants will be possible. The final structures need to form elastic tubes and the implant should not allow crystal formation from urine or harbour local infections. The outer layer of tissue i.e. muscle layer should be lined in the lumenal surface by an intact, differentiating sheet of urothelium.

Support materials usually are resorbable polymers and cross linked collagen sponges. Isolated urothelial cells cultured on collagen sponges formed differentiated sheets of urothelium over the surface of the material, with minimum tendency to promote crystal deposition. Urothelial and bladder muscle cells seeded on to PGA scaffolds formed urothelium like vascularised bilayered tissues when implanted into rabbits.

Recently this technique has been applied to the tissue engineering of a functional bladder in dogs using a fibrous PGA polymer base, shaped into a bladder and coated in PLGA (poly lactic-co-glycolic acid) 50 : 50 co-polymer. Muscle cells were seeded onto the outer surface of the bladder after which the lumenal surface was coated with pre-cultured urothelial cells, prior to implantation. Implanted bladders achieved near-normal performance and maintained this for upto 11 months.

11.4 TISSUE ENGINEERED PERIPHERAL NERVE IMPLANTS

Peripheral nerve injury is a common consequences of trauma and tumour resection surgery, with hundreds of thousands of reconstructive operations performed each year. Two criteria can be used to repair of peripheral nerve injuries. (1) Regeneration of axons should be done from their sprouting at the proximal stump to where they rejoin the degenerating distal stump on the far side of the defect. (2) axonal regeneration across the nerve defect must be as fast as possible. Prolonged delay of reinnervation leads to irreversible muscle atrophy.

Nerve guidance was achieved using silicone conduits attached between the nerve stumps and also reported using tubes of bioresorbable materials such as PLGA co-polymer, collagen type and polymerised hyaluronan. The tube walls provide gross guidance, but no spatial cues are available to axons or Schwann cells away from the tube wall. Support for neurite outgrowth into the tube lumen has been provided by collagen and fibrin, with or without added growth factors.

Initial outgrowth for short distances from the proximal stump can be rapid, but for gaps of > 10-15 mm the process can slow or stop, probably due to a lack of supporting growth factors and Schwann cells. Continued growth can be achieved by adding purified growth factors or by seeding implants with Schwann cells.

QUESTION

1. What is tissue engineering? Discuss its significance.

Index